ASIA, EUROPE, AND THE EMERGENCE OF MODERN SCIENCE

ASIA, EUROPE, AND THE EMERGENCE OF MODERN SCIENCE

KNOWLEDGE CROSSING BOUNDARIES

Edited by

Arun Bala

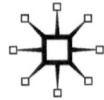

ASIA, EUROPE, AND THE EMERGENCE OF MODERN SCIENCE
Copyright © Arun Bala, 2012.

First published in 2012 by
PALGRAVE MACMILLAN®
in the United States—a division of St. Martin's Press LLC,
175 Fifth Avenue, New York, NY 10010.

Where this book is distributed in the UK, Europe and the rest of the world,
this is by Palgrave Macmillan, a division of Macmillan Publishers Limited,
registered in England, company number 785998, of Houndmills,
Basingstoke, Hampshire RG21 6XS.

Palgrave Macmillan is the global academic imprint of the above companies
and has companies and representatives throughout the world.

Palgrave® and Macmillan® are registered trademarks in the United States,
the United Kingdom, Europe and other countries.

ISBN: 978–1–137–03172–3

Library of Congress Cataloging-in-Publication Data

Asia, Europe, and the emergence of modern science : knowledge
crossing boundaries / edited by Arun Bala.
      p. cm.
   ISBN 978–1–137–03172–3 (alk. paper)
      1. Science—Asia. 2. Science—Europe. 3. Science and civilization.
   I. Bala, Arun.

Q127.A65A85 2012
509.5—dc23   2012003128

A catalogue record of the book is available from the British Library.

Design by Newgen Imaging Systems (P) Ltd., Chennai, India.

First edition: August 2012

CONTENTS

Part III   Forging New Knowledge

Figures

ACKNOWLEDGMENTS

It is my privilege and pleasure to acknowledge those who made this volume possible. I wish to begin by expressing my deepest indebtedness to the director of the Institute of Southeast Asian Studies (ISEAS) in Singapore, Ambassador K. Kesavapany, for his original vision in accepting this project, and his unswerving support afterward, despite the seeming remoteness of history and philosophy of science from the institute's mainstream concerns in politics, economics, and culture. Without his endorsement, the conference that brought together the chapter writers in this volume would not have been possible. I also wish to express my heartfelt gratitude to Dr. Surin Pitsuwan, Secretary-General of the Association of Southeast Asian Nations (ASEAN), who gave the keynote address at the conference, and emphasized the significance of the intercultural history of science for promoting harmony and peace in Southeast Asia as a region shaped by the same civilizations—Chinese, European, Islamic, and Indian—that played major roles in shaping the history of science. I would also like to thank M. Rajaretnam, who convened meetings of the Asian Dialogue Society in cities as diverse as Barcelona, Bangkok, Bali, and Beijing that furnished me with opportunities to interact with leading scholars, businessmen, and political leaders on the significance of intercultural dialogue in a globalizing world. Four friends in particular have served as intellectual sounding boards at various points—Arun Mahizhnan, Jorgen Orstrom Moller, Vijay Kumar Sethi and Geoffrey Wade—and I wish to record my heartfelt appreciation for their insights and encouragement. I also wish to thank Christopher Chappell for being such a wonderful editor by making everything easy, and Sharmila Thiruchelvam and Seetha Sivalingam for assisting in the editing of this work and the zeal that they brought to bear on this task. Finally, I wish to express my sincere appreciation to Selvarani, my wife and partner in this project, for her unstinting enthusiasm and sustained patience over its long period of gestation and fruition.

INTRODUCTION

Arun Bala

Joseph Needham's pioneering studies *Science and Civilization in China* that began more than five decades ago has spawned a vast body of literature that looks not only at the contributions of Chinese science and technology to modern science, but also similar contributions made by other traditions of science—in particular the Indian and Arabic-Muslim. Such accommodation of Asian traditions of science into modern science raises a whole host of historical, epistemological, and sociological concerns that have yet to be systematically integrated into mainstream science studies. While the significance of Needham's contributions has been acknowledged, there has been a tendency to insulate his studies, and others inspired by his works, from mainstream history, philosophy, and sociology of science. This has been noted by Mark Elvin in his introduction to the 2004 volume of the series *Science and Civilization in China* in which Needham lays out his final conclusions and reflections on his monumental work. Elvin writes:

> What is hard to come to terms with, almost half a century after the appearance of the first volume in 1954, is the limited assimilation of Needham's work into the bloodstream of the history of science in general; that is, outside the half-occluded universe of East Asian specialists and a handful of experts sensitive to the decisive contributions of comparisons. For these to be useful, there has to be enough in common between the two domains to make comparisons and contrasts relevant, and enough differences to make such juxtapositions reveal critically distinctive aspects of one or the other.[1]

Yet Elvin himself seems to fail to see that this problem of the marginalization of Needham, and those who followed him, is largely traceable to the Eurocentric conception of the history of science that informs mainstream science studies. The mainstream view is more concerned with the modernist and postmodern debate in science studies typified by the Popper versus Kuhn controversy.[2] It is significant that both Popper and Kuhn, despite profound differences in their epistemological positions, share the Eurocentric presumption that modern science is historically rooted only in the ancient Greek heritage of science. In his essay "Science: Conjectures and Refutations" Popper argues that the discovery of the critical method that is so essential for modern science began with the Greeks—more specifically with Thales and the Presocratics who followed him. He writes:

> I think that it was Thales who founded the new tradition of freedom— based upon a new relation between master and pupil—he seems to have been able to tolerate criticism. And what's more, he seems to have created the tradition that one ought to tolerate criticism...It was a momentous innovation. It meant a break with the dogmatic tradition which permits only one school doctrine, and the introduction in its place of a tradition that admits a plurality of doctrines which all try to approach the truth by means of critical discussion. It thus leads, almost by necessity, to the realization that our attempts to see and to find the truth are not final, but open to improvement; that our knowledge, our doctrine, is conjectural; that it consists of guesses, of hypotheses, rather than of final and certain truths; and that criticism and discussion are our only means of getting nearer to the truth.[3]

Like Popper, Kuhn also seems to embrace a Eurocentric conception of the history of modern science. He writes:

> Every civilization of which we have records has possessed a technology, an art, a religion, a political system, laws, and so on. In many cases those facets of civilizations have been as developed as our own. But only the civilizations that descended from Hellenic Greece have possessed more than the most rudimentary science. The bulk of scientific knowledge is a product of Europe in the last four centuries. No other place and time has supported the very special communities from which scientific productivity comes.[4]

In this context it is interesting to contrast the views of Popper and Kuhn with one of the early philosophers of modern science—Francis Bacon. He is often deemed to have pioneered the experimental method of science that inspired Newton and the Royal Society in seventeenth-century England. Instead of seeing the Greeks as the inspiration for science, Bacon saw the Aristotelian heritage as an obstacle to the emergence of modern science, and his *Novum Organum* where he developed his famous inductive and experimental methodology was consciously set against the contemplative and logicist orientation of Aristotle's *Organon*. It is also noteworthy that in another study, *New Atlantis*, published six years later in 1626, Bacon imaginatively describes a fictional civilization located in the Pacific Ocean between the Americas and Japan in which the sciences and arts were systematically cultivated. He postulated that the population of New Atlantis was itself the result of the fact that about

3000 years ago navigation was far more advanced than it had been since. At that time the Phoenicians, Egyptians, Carthaginians, and Chinese had large fleets and traded with each other—but Persians, Chaldeans, and Arabians also traveled on these ships. Even in the Americas there were great civilizations in Peru and Mexico who had ships that crossed the Atlantic and Pacific oceans. However, these civilizations (especially in the Americas) were destroyed by a massive inundation and the people declined to the more simple conditions in which they were found in Bacon's time. New Atlantis not only became isolated from the rest of the world but also set out to insulate itself from the general catastrophe that was overtaking the world as a whole.

However, New Atlanteans while isolating themselves nevertheless sent out regular expeditions to learn from others about new discoveries and inventions. The most significant element in New Atlantis was Solomon's House—a huge laboratory and library for storing and cultivating scientific knowledge. From Solomon's House are sent emissaries to other countries and nations of the world to collect knowledge of diverse sorts. Clearly Bacon's utopian essay describes the social organization of science that became an important factor in the growth of modern science and the applied technology associated with it. It was also to become an important influence on the founding of scientific societies in Europe. One significant element in his work is the crucial contribution that the collection of knowledge from diverse parts of the world can make to the advancement of science. The history of modern science often creates the impression that science developed within Europe with little input from the outside. Yet in this influential study it is evident that Bacon did not underestimate the contributions that other cultures could make to the growth of the sciences.

This raises a series of important questions: To what extent did a dialogue of civilizations—a dialogue recommended by Bacon at the dawn of science—play a role in the rise and growth of modern science? The chapters in this volume began as contributions to a conference that brought together writers to see what common issues, problems, themes, and questions can be identified for the history, philosophy, and sociology of science once we begin to take seriously the dialogical encounters that enriched modern science. Given this orientation, the book naturally divides into three sections. The first section addresses historical issues connected with the flow of ideas from Asian traditions of science into Europe that promoted the rise of modern science; the second with epistemological issues associated with the discovery and justification of the knowledge thus produced; and the third with the sociology of knowledge associated with research and pedagogy orientations.

The historical section brings together five chapters that deal loosely with a set of interconnected issues. It begins with John Hobson's attempt to challenge what he terms a neo-Eurocentric construction of history that has increasingly come to displace classical Eurocentrism. The latter was the tendency until quite recently to minimize, marginalize, or ignore the scientific and technological contributions of Asian civilizations to modern science in Europe. As this became increasingly difficult to sustain in the face of accumulating historical evidence to the contrary, neo-Eurocentrists adopted a new strategy. They acknowledged the borrowings of non-Western resource portfolios (ideas, institutions, and technologies) in the rise of the West, but argued that it was the unique cosmopolitanism, tolerance, and cultural openness of the West that made it possible for it, and no other civilization, to draw upon diverse cultures to produce modernity. Hobson's chapter is an attempt to investigate how we may go about countering this more-nuanced version of Eurocentrism, that Europe was particularly situated to engage in dialogical encounters with other civilizations because of certain essential unique values that it carried. Although he offers no conclusive solution, he does reveal how earlier attempts to counter Eurocentrism by piling up evidence of influences from Asia may not be sufficient to transcend neo-Eurocentric conceptions of the history of science.

George Gheverghese Joseph's chapter is another attempt to counter historical Eurocentrism. He notes that often claims that Europe was influenced by discoveries made in other cultures are dismissed on the grounds that the same discoveries were later made independently in Europe. He examines different methodological approaches to determine whether parallel developments of ideas in two cultures are to be taken to be independent developments or the outcome of dialogical encounters. In particular, he shows how such methodologies can be used to evaluate the claim that the invention of the calculus, and the "passage to infinity" that led to the discovery of the infinite series expansions of trigonometric functions in Europe in the seventeenth century, could be used to test the hypothesis that they were inspired by earlier discoveries in Indian mathematics.

Michal Kokowski addresses an equally significant concern raised by dialogical histories. Often such histories are also motivated by the desire to disparage the achievements of European scientists by putting into question their originality. Taking the example of Nicholas Copernicus, Kokowski considers the argument made by some scholars of Arabic science that Copernicus could not be considered an original revolutionary thinker because he simply adopted the best methods and results of ancient Hellenic and Hellenistic science, as well as medieval Arabic science, without contributing any substantial new ideas. Indeed some scholars who see his thought as a natural development of medieval Arabic astronomy have claimed that he is simply "a European satellite of the Maragha astronomers." Kokowski develops the concept of a scientific (r)evolution that combines the notions of evolution and revolution. He uses it to explain why even after we acknowledge dialogical encounters to have influenced the discovery of a seminal scientific thinker, whose ideas can be seen as an evolutionary development from such encounters, it is still possible to view the new development as revolutionary. Kokowski maintains that seeing Copernicus as promoting a scientific (r)evolution makes it possible to treat his achievement as revolutionary even if it can be interpreted as an evolution grounded upon the achievements of Arabic science.

Richard Arthur also addresses the problem of allowing for originality, while acknowledging influence, and like Kokowski is motivated to introduce a new concept to defend his position—in his case the notion of "epistemic vectors." His illustrative example is also drawn from the influence of Arabic-Islamic ideas on European thinkers. He argues that the influence of earlier thinkers need not occur in the form of a direct transmission of ideas—it could be the result of an earlier dialogical encounter that provided the context for the later independent development of parallel ideas. He suggests that influences from Arabic culture in an earlier period, could have placed European thinkers in similarly constrained historically and socially conditioned trends of thought—epistemic vectors—that led to independent future parallel developments in their thinking. He illustrates this by showing how the doctrine of Occasionalism could have arisen in seventeenth-century Europe without the direct influence of earlier similar doctrines in Islamic Ash'arite theology. Cartesian Occasionalists and Islamic Ash'arites could have developed the same critiques against the necessity of cause-effect linkages assumed in ancient Greek thought because European knowledge of Greek sources had been refracted through the contribution of Islamic thinkers.

The last chapter in the historical section by Roddam Narasimha argues that the great progress made in modern mathematics in the seventeenth-century scientific revolution has to be explained as the outcome of a dialogical encounter that brought about the cross-fertilization of Indian and ancient Greek mathematical modes of inquiry, or epistemologies. In particular he maintains that the transmission of certain Indian epistemological ideas associated with the notion of *yukti*, or skillful techniques, in mathematics, and their combination with the notion of proof in Europe inherited from the ancient Greeks, played a key role in the rise of early-modern mathematics in Europe. He also highlights the significance of such transmissions by suggesting that they also contributed to the dramatic advances in science made in Europe over the last four centuries.

The second section brings together five chapters that address epistemological issues associated with the dialogical perspective. The first chapter by James Brown sets out to define how we should approach the dialogical encounters of Asian traditional knowledge and modern science. He highlights the limits of the modernist approach of rejecting traditional knowledge as pseudoscience, the postmodern embracing of cognitive relativism that treats modern science as on par with traditional knowledge, and the acknowledgment of traditional knowledge as often technologically viable but theoretically naive. He thinks that none of these approaches engage the theoretical beliefs of traditional knowers, and attempts to offer an approach to how we can do this without relinquishing acceptable conceptions of scientific rationality. In the process, he proposes a more-nuanced mode of inquiry, or epistemology, able to integrate viable theoretical ideas and practices from premodern traditions into modern science and technology.

Where Brown attempts to offer an "objectivist" approach to taking Eastern traditional knowledge seriously, Arun Bala develops a neo-Lakatosian notion of research programs to explicate how theoretical ideas of Asian traditions could have been incorporated into modern science. He illustrates this with the Copernican revolution. He shows how accommodating the sun-centered vision of the universe required major revisions in other areas of knowledge—mathematics, optics, mechanics, and cosmology. He demonstrates how these revisions were made possible by incorporating optical ideas from Arabic science, mathematical ideas from Indian science, and astronomical ideas from Chinese science. He argues that a neo-Lakatosian model of the growth of scientific knowledge can show how research programs from different cultures, for example, Arabic optics, Indian mathematics, and Chinese astronomy, were dialogically combined to advance scientific knowledge.

While Brown and Bala are concerned with showing the relevance of theoretical ideas in Asian traditions of science for modern science, Ali Paya rejects the notion that there are theoretical ideas in science that can be deemed distinctively Islamic or indigenous (i.e., specific to a particular culture). Such a position presumes that adopting a religion or a particular tradition or form of life can be a precondition for accepting a scientific or theoretical idea, and that there can be a tradition of so-called "Islamic science" or "religious or indigenous science." By examining several projects for producing "Islamic science" in a number of Muslim countries in the recent past, he shows why all such projects, not withstanding pious intentions behind them, are misguided and doomed to failure. Instead he argues that science is universal even if its growth is mediated by dialogical encounters across cultures. However, he also maintains that this does not preclude the possibility of "Islamic technologies"—that is, technologies motivated by Islamic values and goals.

Anjam Khursheed's chapter looks at the epistemological orientations of cultures that promote or hinder the growth of scientific and engineering knowledge. He finds that there are striking parallels in Asian and European cultures in their periods of sterility and creativity. Periods of sterility were periods of isolation that were often associated with religious revival movements that sought to marginalize and suppress foreign influences by appealing to the inherent supposed superiority of some religious authority. However, periods of creativity were periods of cultural openness characterized by a forward-looking inclusive worldview, where the achievements of foreign cultures were embraced and incorporated. Hence, Khursheed concludes, periods of scientific creativity in Asia and Europe were also times when these cultures were open and receptive to dialogical encounters with other cultures, and that periods of scientific sterility coincided with epochs of religious dogmatism and revivalist movements that closed off such dialogue.

The final chapter in the section by Andrew Brennan critiques Richard Nisbett's thesis that the emergence of modern science in Europe rather than in China can be explained as the outcome of cognitive processes shaped by the historical and environmental circumstances in which European thought emerged with the Greeks. According to Nisbett, Chinese civilization emerged with agriculturalists in China and produced a "planter mentality," since people needed to work in close cooperation with each other. The planter mentality was deferential to authority, averse to contentious debate, and inclined to holistic thinking quite tolerant of inconsistency. By contrast, the earliest inhabitants of Greece were dependent on fishing, hunting, and trade that encouraged what Nisbett terms a "pirate mentality." This promoted adversarial questioning of authority, and an inclination for atomistic, analytical thinking with a high regard for logical consistency. Nisbett concludes that the Greek pirate mentality is more conducive to the development of modern scientific thinking with which it has a close affinity. Brennan questions whether such cultural differences can be explained simply in terms of the environments that conditioned early European and Chinese thinking. Rather he argues that both kinds of mentalities can be found in all cultures, but that the greater proportion of Europeans having pirate mentalities in contrast to Asians is the product of the long dominance of scientific thinking there, and not the cause of science.

The final section brings together four chapters that address the sociocultural reorientation needed in research and education once we adopt the dialogical perspective. It begins with Keekok Lee's attempt to understand the relevance of classical Chinese medicine for future science. She argues that the causal paradigm behind the modern revolution in science that developed in seventeenth-century Europe is coming to be supplanted by a "post-postmodern concept of cause" with a holistic metaphysics that is non-Humean, nonlinear, and nonreductionist in character. She defends the notion that this kind of "ecosystemic science" is similar to the theory and practice of traditional Chinese medicine. This suggests, she argues, that a dialogical encounter with the alternative ecosystemic paradigm of classical Chinese medicine can provide a fruitful way of conceptualizing the world and coping with many complex environmental problems we will face in the coming decades of this century.

The chapter by Donald Wiebe also points to the important contributions that Eastern thought can make to future science—particularly the scientific study of religion. Wiebe is intent to repudiate the notion that the scientific study of religion is necessarily Eurocentric, since it imposes an essentially Western perspective on Eastern religious ideas, and thereby serves the hegemonic agenda of Western imperialism by violently suppressing non-Western ways of life, forms of knowledge, and constructions of reality. He rejects such accounts on the grounds that such a study of religion can actually draw upon dialogical encounters with Eastern thought, because certain strands of Eastern thought already carry, albeit in germinal form, notions that can contribute to a scientific approach to religion. Hence, like Lee, he also finds that Eastern thought can contribute to the growth of future scientific knowledge.

While the chapters by Lee and Wiebe address the research implication of dialogical approaches, the last two chapters by S. Gopinathan and Victor Savage are more concerned with its implications for education. Gopinathan argues that schooling is a culture-specific activity and that more effective learning can result if educators and teachers acknowledge and respond to cultural contexts and traditions. Historically, Asian cultures had well-established structures for knowledge generation and dissemination—for example, Nalanda University in ancient India, the Chinese imperial examination system, and madrassas across the Islamic world. These have largely been supplanted after industrialization, modernization, and colonial reach by systems of mass education that have internalized Western models of knowledge production and inculcation. His chapter examines the implications of making a culture-sensitive turn, one able to promote more egalitarian dialogical encounters between modern and earlier Asian traditions of pedagogy, for teacher preparation.

The final chapter by Victor Savage also addresses the relevance of an education that can promote a dialogue between traditional knowledge and Western ways of thinking. He argues that Southeast Asian concepts of nature and cosmology carry "folk science" knowledge in their cosmic representations and worldviews that are particularly relevant to build a sustainable future. He therefore concludes that we need a pedagogy that integrates science with traditional knowledge, and Western ways of thinking with Asian cultural perspectives, especially in the area of environmental education.

Taken together, the chapters in this volume serve to fill a lacuna in science studies. Much of contemporary history, philosophy, and sociology of science is conducted without taking into account the wider global context in which modern science emerged. If that emergence can be taken to have begun with the publication of the *Revolutionibus* of Copernicus in 1543, then this event was preceded by a half-century of voyages of discovery that led Europeans to the Americas, India, Southeast Asia, and East Asia. How did this contact with the wider world shape the rise of modern science? What contributions did these cultures beyond Europe make to modern science? How would acknowledging such contributions require us to rethink not only the historical roots of science, but also its philosophical basis and the social context of its origination? How would it also require us to rethink the role of diverse cultural reservoirs of knowledge in advancing future science? Although the writers in this volume provide no consistent or definitive answers to these questions, they nevertheless open the door to a new kind of dialogue of civilizations for the history, philosophy, and sociology of science.

END NOTES

1. Mark Elvin's *Introduction* in Joseph Needham, *Science and Civilization in China* (Cambridge: Cambridge University Press, 2004), p. xxv.
2. For a recent study of the continuing significance of this debate, see Steve Fuller, *Kuhn vs. Popper: The Struggle for the Soul of Science* (New York: Columbia University Press, 2004).
3. See Karl Popper, "Back to the Presocratics," *Conjectures & Refutations: The Growth of Scientific Knowledge* (London: Routledge ,1963).
4. Thomas Kuhn, *The Structure of Scientific Revolutions* (Chicago, IL: Chicago University Press, 1970), pp. 167–168.

BIBLIOGRAPHY

Kuhn, Thomas. 1970. *The Structure of Scientific Revolutions*. Chicago, IL: Chicago University Press.

Needham, Joseph. 2004. *Science and Civilization in China,* vol. 7. Cambridge: Cambridge University Press.

Popper, Karl. 1963. *Conjectures & Refutations: The Growth of Scientific Knowledge*. London: Routledge.

Steve Fuller. 2004. *Kuhn vs. Popper: The Struggle for the Soul of Science*. New York: Columbia University Press.

I
Toward Dialogical History

1

GLOBAL DIALOGICAL
HISTORY AND THE CHALLENGE
OF NEO-EUROCENTRISM

John M. Hobson

The worst thing ethically and politically is to let [Eurocentric] separatism simply go on, without understanding the opposite of separatism, which is connectedness...What I am interested in is how all these things work together. That seems to me to be the great task—to connect them all together—to understand wholes rather than bits of wholes...In a wonderful phrase, Disraeli asks, "Arabs, what are they?" and answers: "They're just Jews on horseback." So underlying this separation is also an amalgamation of some kind.

Edward W. Said[1]

It has become the fashion to level the charge of Eurocentricity at the West for ignoring our debt to the achievements of other civilizations. Yet while fully acknowledging this debt, we must still ask why the West, after the end of the Middle Ages, so rapidly overtook the great civilizations of the East.

Ernst Gombrich[2]

When talking of the Middle Ages...historians of science...tend to talk of the "transmission" of Greek and Arabic learning to the West, or about how the West "received" [these ideas]...all painting a passive picture of the process...[But Westerners] *actively* sought [Arabic learning]...The West was to learn much from Islamic scholars, philosophers, and scientists like Al-Farabi, Avicenna, and Averröes, and even learned to respect "chivalric heroes like Saladin." It is worth pondering that there was a period in Islamic civilization when it was open to outside influences in ways similar to the West. Early Islamic civilization derived much from the Greeks, Romans, Indians, and Persians. But crucially, at a certain date, it rejected the "foreign sciences," which led to intellectual stagnation and decay.

Ibn Warraq[3]

INTRODUCTION

In the last few years the idea of the "dialogue of civilizations" has begun to permeate into the study of non-Eurocentric global history. Its prime rationale is to challenge two interrelated approaches—Samuel Huntington's idea of the "clash of civilizations" and Eurocentric world history. Let me discuss each in turn. Huntington, of course, characterized intercivilizational relations as inherently conflictual, on the basis that civilizations are self-contained entities that have their own unique cultures that are, in turn, incommensurable with those of other civilizations.[4] This culminates in his view that the meeting point between civilizations can be likened to tectonic platelike fault lines, which abrade to produce or generate violent and bloody conflict. It is helpful here to differentiate two forms of civilizational analyses—substantialist and processual/relational. A substantialist approach is essentialist, wherein civilizations are thought to display essential characteristics that are largely static or unchanging. By contrast, a relational approach conceives of civilizations as sets of social practices such that their boundaries are written or drawn and redrawn over time through intercivilizational interactions.[5] However, while Huntington might balk at being placed in the substantialist category, since he does, in fact, argue that civilizations change over time,[6] nevertheless the logic of his position remains otherwise, given that the traditional and primordial cultural/religious values that he focuses upon are by definition unchanging. Moreover, the second defining feature of substantialism seals his position within this category. This concerns the point that substantialist accounts view the reproduction of civilizations as endogenously generated. By contrast a relational approach—as the term properly implies—insists that civilizations are shaped and constituted, reshaped and reconstituted, through iterated interactions with others around and beyond them.

Huntington's commitment to the idea that civilizations endogenously *self-generate* is also a fundamental aspect of Eurocentric world history. Although some of the ideas upon which Eurocentrism was founded began to emerge in the period following the "discovery of America," it was only during the European Age of Enlightenment that Eurocentrism was really crystallized, as European thinkers set about determining Europe's place in the world on the one hand, and sought to construct a new, modern European identity on the other. Prior to, and even during much of, the eighteenth century, Europeans often recognized that the East and the West were interlinked. But the emergence of Eurocentrism and the concomitant "production of alterity" led to the construction of what I like to call an imaginary *line of civilizational apartheid* that fundamentally separated the East from the West. While Jessica Benjamin usefully refers to this as the process of "splitting" the East and the West apart,[7] I would prefer to call it an "intellectual cleansing," or simply a "purging," of the Eastern Other from the Western Self. This is *the* crucial move, and finds its harshest expression in the writings of the late-nineteenth-century scientific racists who looked to literally "purify" the Western self of Eastern "contamination" sometimes through the extermination of the Eastern races (as in, e.g., Karl Pearson), or through the erecting of prohibitive barriers to nonwhite immigration into the West (as in most notably Lothrop Stoddard).[8] Ultimately this move renders invisible the dialogues of civilizations that mutually constitute them. In turn, having split or "purified" these mutual, hybrid civilizations into distinct and reified, self-constituting entities, Eurocentric thinkers then elevated the Western Self and demoted the Eastern Other. The West was imbued with exclusively progressive or rational characteristics, including liberal democracy, liberal capitalism, individualism, and rational science, all of which ensured that the West would not only make political and economic modernity single-handedly, but would also be the torchbearer of political/economic development in the world. By contrast, the Eastern Other was imbued with all manner of regressive and antithetical properties—including Oriental despotism, agrarian systems of production, collectivism, and irrational mystic religions—all of which ensured that stagnation would be its lot. This culminated in Max Weber's famous distinction between the Western "ethic of world mastery" and the fatalistic Eastern "ethic of passive conformity" to the world. Thus, Western man was elevated to the permanent "proactive subject" of global politics/economics—past, present, and future—standing at the center of all things. Conversely, Eastern "man" was relegated to the peripheral status of global politics' "passive object," languishing on the Other side of an imaginary civilizational frontier, stripped of history and dignity. In this Eurocentric imaginary, then, an intellectual line of civilizational apartheid separates the Western heart of light from the Eastern heart of darkness.

Having constructed Europe as superior and *exceptional* by the early nineteenth century, Romantic thinkers then extrapolated this conception back in time to ancient Greece, thereby painting an ahistorical picture of permanent Aryan Western supremacy. It was, of course, round about this time when the social sciences were emerging. But rather than critique this metanarrative, social scientists unreflexively endogenized it into their theories. Accordingly, they explain Europe's rise by excavating causal variables that allegedly exist only within Europe. This presupposes the Eurocentric endogenous *logic of immanence* through which Europe's rise is self-generated before it subsequently projects its global will-to-power to remake the world in its own image. Thus having extrapolated European supremacy back in time to ancient Greece, Eurocentric thinkers then trace forward world political and economic development through an immanent journey of the Western "Oriental Express," as the train of development passes through a series of European way-stations including feudalism, the commercial/financial revolution, and onto the Renaissance, through the Enlightenment and British industrialization to arrive at the global terminus of history—the Pax Americana/high mass consumption for liberals, and communism for Marxists. Conversely, such a progressive linearity was absent on the Other side of the "civilizational frontier." In the process, the West is granted an "iron law of development," while the East suffers an "iron law of non-development." Accordingly, the Easterners could only passively await the arrival of the Oriental (imperial) Express that, fueled by an *Occidental/Eurocentric Messianism*, steamed across to pick them up, to either graciously deliver them to the emancipatory terminus of history (as for liberals and classical Marxists) or to relentlessly hold them down through exploitation (as for most neo-Marxists).

In recent years a new approach is emerging that confronts both these conceptions of world politics and world history. This approach emphasizes the "dialogues of civilizations" (*global dialogism* from henceforth).[9] And, not surprisingly, it takes issue with Huntington's thesis as well as Eurocentrism (though this in turn is hardly surprising given that Huntington's work is clearly Eurocentric, indeed explicitly so).[10] It shares much in common with other non-Eurocentric analyses,[11] though I think it fair to say that it also marks out its own distinctive intellectual space. Concerning global dialogism's anti-Eurocentric stance, its twin key points for departure lies with its rejection of the European "logic of immanence" and the concomitant notion of European exceptionalism and superiority on the one hand, and the notion that the East and the West are separate and self-constituting on the other. In its place it emphasizes the centrality of dialogical civilizational interactions that in turn generate hybrid civilizational properties, thereby returning the Eastern Other into the heart of the Western Self. In particular, the rise of the West is rewritten to show that Europe did not self-generate according to an exceptional internal script but was fundamentally "enabled" by the transmission of non-Western "resource portfolios" (comprising Eastern ideas, institutions, and technologies).

This focus in turn will bring the East back into the picture of the causes of modernity while simultaneously downgrading the supreme agency—or hyperagency—of the West. However, my feeling is that merely filling in the details of these transmissions is, albeit important, ultimately insufficient in the light of a series of profound challenges that are issued by a nascent neo-Eurocentric paradigm. Revealing these challenges and suggesting at least some solutions to them is my principal concern in this chapter, given my belief that they threaten the very integrity of non-Eurocentric global dialogism.

There are two generic challenges that nascent neo-Eurocentrism poses either directly or indirectly, though these contain within them various further challenges. The first generic problem with global dialogism lies in its tendency to lapse into a latent Occidentalism. This emerges insofar as the West is painted as a *passive beneficiary* of Eastern resource portfolios. At the extreme, this simply inverts the Eurocentric approach that assumes that having made itself, the West then diffused its standard of civilization to the rest of the world so that Eastern civilizations could also develop. This danger is by no means irredeemable given that it is not an inherent problem and it can be rectified. But, doing so requires responding to the second interrelated challenge. This concerns my belief that non-Eurocentric global dialogism is in danger of being outflanked by the emergence of what I will call "neo-Eurocentrism," or "neo-Orientalism." This approach has yet to be fully developed though I anticipate its development in the coming years. My attention was first drawn to this nascent approach, when I entered into a journal debate with Ricardo Duchesne, a self-confessed Eurocentric, in 2006.[12] Duchesne criticized me and other non-Eurocentrics for unfairly representing the Eurocentric approach to global history, claiming that many Eurocentric world historians, especially since about 1990, are sensitive to the point that Europeans often borrowed Eastern "resource portfolios" and that it was this "cosmopolitan" attitude that marked the exceptional quality of Europe and the West. Above all, this focus shifts the theoretical terrain away from pure European inventive genius toward the idea of European creative adaptability.

While I was unconvinced that Eurocentric scholars have effected this kind of move, nevertheless the thrust of Duchesne's argument, and the recent arguments of Ibn Warraq as well as Toby Huff, combine to offer a glimpse of a new neo-Eurocentric landscape.[13] And here I want to signal a point that I discuss later. For of the various challenges that neo-Eurocentrism offers up, the most profound one concerns the point that it is able to offer an *explanation* of the rise of the West that is missing within global dialogism. Overall, the two generic challenges or problems that are mentioned here can only be adequately met with the development of a non-Eurocentric theoretical explanation of the rise of the West. And, as we shall see, this, I confess, is an enormously difficult challenge.

The chapter proceeds as follows. The first section provides a sketch of the principal features of neo-Eurocentrism's explanation of the rise of the West and how this shifts the terms of the debate back onto Eurocentric terrain. The second section then reveals how non-Eurocentrism can account for European adaptability but cannot adequately explain the rise of the West. Finally, the third section outlines the conundrums that face non-Eurocentrism when trying to build an explanation and closes by suggesting one possible response. Even so, I concede that this only gets us half the way and that more thinking needs to be done.

A NEO-EUROCENTRIC APPROACH?
SHIFTING THE TERMS OF DEBATE
AGAINST GLOBAL DIALOGISM

I would argue that neo-eurocentrism has been in the offing for a number of years, if not decades, but that it is only recently that it has begun to take a more crystallized form. The basic principle is well articulated by Duchesne:

> A distinctive trait shown by Europeans was precisely their willingness to imitate inventions made by foreigners, in contrast to the Chinese who ceased to be as inventive after the Sung era, and showed little enthusiasm for outside ideas and inventions... A major secret of European creativeness is precisely its multicultural inheritance and its wider geographical linkages with the peoples of the world.[14]

In the past such a conception has certainly existed but it was not developed beyond occasional statements that litter the Eurocentric literature. Duchesne takes inspiration from Carlo Cipolla's 1976 book,[15] though other examples can be marshaled. Francis Oakley asserted:

> What made [the West] extraordinary was less the capacity to invent than the readiness to learn from others, the willingness to imitate, the ability to take over tools or techniques discovered in other parts of the world, to raise them to a higher level of efficiency, to exploit them for different ends and with a far greater degree of intensity.[16]

And no less an authority as Toby Huff, in his study of the causes of the European scientific revolution, insists:

> Modern science is the product of intercivilizational encounters, including, but not limited to, the interaction between Arabs, Muslims, and Christians, but also other "dialogues between the living and the dead" involving Greeks, Arabs, and northern Europeans. Indeed, some would say that it was the Greek heritage of intellectual thought...that set the course for intellectual development in the West ever after. One does not have to subscribe to such a view to recognize the great importance of the Greek tradition to Western science. The larger point is, however, that modern science is the end product of several such sustained intercivilizational encounters over the centuries.[17]

This nascent approach has the capacity to swallow whole the many insights concerning the transmission of Eastern inventions that non-Eurocentrics point to. That is, no matter how many instances of borrowing the non-Eurocentrics unearth, the neo-Eurocentric reply is that these matter not. What does matter, however, is precisely the ability of the Europeans to assimilate and adapt these Eastern resources to higher ends. In the process, this brings to the surface the key claim made originally by Cipolla. Thus while Cipolla insists that the Europeans developed an original inventiveness from the twelfth century on, crucially he also asserts that on the (albeit few) occasions "when Europe absorbed new ideas from outside, it did not do so in a purely passive and imitative manner, but often adapted them to local conditions or to new uses with distinct elements of originality."[18] The quote is interesting in that on the one hand it wants to suggest that Europe's key to success lay in its ability to adapt or assimilate others' inventions to higher ends, but it is simultaneously grudging in terms of recognizing the many Eastern contributions.

My argument here is not to counter this with a renewed drive to unearth evermore Eastern borrowings (important though they have been), for this would mean entering into a quantitative numbers game that I believe would be to miss the "qualitative" challenge being issued here. For I believe that we are on the cusp of a new neo-Eurocentric paradigm that poses a fresh challenge to global dialogism. In essence, the new challenge is of a picture of Europe that is no longer the master of invention and the creator of everything. Rather, the picture that is emerging is one of a Europe that is superior and exceptional precisely because of its ability to imitate and borrow from others before subsequently adapting these to higher ends. This in turn means that the target of non-Eurocentrism is now morphing. For the old view of a Europe that developed through an endogenous logic of immanence is now being sidelined in favor of an approach that recognizes and subsumes the insight of the external, Eastern input into the rise of the West. At the extreme, the insights of dialogism could even be used to enhance the neo-Eurocentric approach, since what matters is not the number of Eastern inventions that have been borrowed but the point that Europe was able to work with them and assimilate them to higher ends.

In recent years neo-Eurocentrism (though the term does not currently exist within the literature) has begun to take a more crystallized form. I believe that this is what Ricardo Duchense is driving at, and is at least implicit within Toby Huff's 2003 book. But the clearest and most sustained articulation can be found in Ibn Warraq's 2007 book, *Defending the West*. Arranged as a full frontal attack on Said's critique of Orientalism/Eurocentrism, Warraq devotes the majority of the book (Part 2) to considering the "three golden threads," or the "three tutelary guiding lights," of Western civilization within a historical framework, comparing Europe with the Islamic Middle East.[19] These three threads comprise rationalism, universalism, and self-criticism. It is important to note that like Edward Said, Warraq conflates the Arabic Middle East with the East in general; clearly a problematic conflation for the purposes of this investigation. But that aside, Warraq's key claim is that the West has, from ancient Greek times, been engaged in a quest for the truth and values knowledge for its own sake. This is constantly contrasted with the Islamic Middle East, where knowledge is perverted to fulfilling the needs of religious authorities so that, accordingly, there is no notion of truth outside the stultifying sphere of Islamic revelation. It is the desire to know the truth that gives rise to a major thrust of his argument—that Europeans have from the beginning been deeply *curious* about the world and have been hungry to learn about other cultures, languages, and literatures. By contrast, this intellectual curiosity and thirst for objective knowledge is denied in Islam, which apparently has shown only contempt for foreign languages and literature. "Even after eight centuries of Muslim presence in Spain, we know of only a single document that reveals any Muslim interest in a European language."[20]

It is tempting here to offer counters, such as the point that the Muslims showed little interest in Europe because prior to 1500 Europe had little to offer that the Muslims did not already know. Or that clearly some Muslims had gone through the trouble of learning Greek not least so as to translate the ancient Greek texts; or that what is particularly striking about the Islamic world has been its intense curiosity for non-Islamic sources of knowledge, which is precisely why the seventh Abbasid caliph, al-Ma'mūn, founded the "House of Wisdom" (*Bayt al-Hikmah*) in Baghdad where *inter alia* Greek works—especially those of Ptolemy, Archimedes, and Euclid—as well as Persian, Indian, and Chinese—were translated into Arabic. But I shall resist temptation and confine my task to merely reporting Warraq's basic argument.

Tracing back to ancient Greece and then forward, he claims that the Greeks were open to new ideas from outside—ideas garnered from Vedic culture as well as the Near East and Egypt. But critically, it is the capacity to put these ideas to higher ends that marks the creative or adaptive genius of Greece. In this neo-Eurocentric approach what matters most is the ability to imitate and adapt—a notion that flips traditional Eurocentrism on its head, since such "qualities" were originally reserved only for the inferior Eastern cultures. Now, it turns out, *imitation* (and adaptation) is the mark of a higher culture. Here he quotes from Walter Burkert, arguing that "through acculturation something new may arise; and although Greeks had been on the receiving side for a long time, there is no doubt that the result is Greek. It is Greek art and culture that has become classical, and Greek literature that has become world literature."[21] This in effect places the idea of *emergent* properties at the centre of that which makes Europe unique.

In this way, I would suggest, the old markers of Eurocentric world history—the European logics of "immanence" and "inventive exceptionalism'—are in the process of being replaced with the neo-Eurocentric markers of the West: the logics of "emergence" and "adaptive/imitative exceptionalism." In turn, this leads to a new picture that potentially out-flanks the non-Eurocentric dialogical critique of Eurocentrism. Indeed, if one probes further the implications are profound. For it brings into question much of the critique of Eurocentrism. It means, for example, that viewing the East and the West as *entirely* separate is no longer the hallmark of Eurocentrism. It similarly problematizes the assumption that Eurocentrism invests Europe with a unique inventiveness, since it con-cedes (albeit somewhat grudgingly) that significant inventions have been made outside of the West. Nevertheless there is still very much in evi-dence a continuing commitment to the idea of Western exceptionalism.

It might be replied that this commitment to Western exceptional-ism still betrays an unacceptable bias that underpins neo-Eurocentrism. And given this, non-Eurocentrics might rest a little easier, and continue their search for arguments that seek to deconstruct this myth and replace this meta-hierarchy with a more democratic conception wherein the East and the West are placed on a similar inventive or creative footing. But precisely at this point, neo-Eurocentrism implicitly offers up a double-whammy for the non-Eurocentric dialogical approach. For as I indi-cated earlier, non-Eurocentric dialogism is being outflanked by virtue of the fact that neo-Eurocentrism can offer an *explanation* of the rise of the West that, I would argue, is currently missing within the non-Eurocentric global dialogical approach. And until this is rectified, the non-Eurocentric approach may be likened to the proverbial Titanic prior to it striking an iceberg on its maiden voyage. Of course, this is perhaps unnecessarily hyperbolic given that I believe that the non-Eurocentric ship can be saved, but it serves not least to sound the ship's alarm bell well in advance. Before considering how the ship might be diverted from its potential fate, it is worth briefly surveying the outlines of the iceberg.

SURVEYING THE NEO-EUROCENTRIC CHALLENGE

Here we confront a paradox: that global dialogical and non-Eurocentric scholars more generally have an *implicit account* of Europe's adaptabil-ity but, unlike Eurocentrism/neo-Eurocentrism, they have *no theory* of Europe's adaptability. Regarding the implicit account of European adaptability, much of the non-Eurocentric literature can be subdivided into various approaches, which I shall briefly survey.

The first variant might be termed "global structuralism" and is most forcefully advocated by Janet Abu-Lughod and Andre Gunder Frank. For Frank, the key moment was the post-1762 downturn in the global economy (opening up a Kondratieff B cycle). His argument boils down to the propositions that the Europeans adapted to the global downturn by developing labor-saving technologies such as the steam engine to reduce costs given the problem of high-wage labor, whereas the Eastern economies, which enjoyed low-wage labor, had no need to adapt in this way.[22] Abu-Lughod also views changes in the global economy as proving vital. For her, the West was able to move ahead and was able to adapt at a point when the East was in temporary disarray at precisely the wrong moment—that "the Fall of the East preceded the rise of the West."[23] The key development was the Black Death of the mid-fourteenth century, which disrupted the major trade routes that were governed over by Eastern actors and into the subsequent vacuum that was created by the withdrawal of the Easterners from the global economy the Europeans poured in, with the rest being (Western) history.

A second variant is found in the "contingency approach," advocated most forcefully by Kenneth Pomeranz and James Blaut. Blaut's argument is that before 1492 Europe and Asia enjoyed similar levels of development and showed no differences in institutional capacity or cultural outlook. What led on to the divergence thereafter was that the Europeans benefited from a particularly fortuitous contingency: that Europe was closer to the Americas than was any other continent.[24] Thus having fortuitously stumbled on the Americas where the Europeans no less fortuitously stumbled on gold and silver, their subsequent plundering of this bullion supplied the capital necessary to stimulate capitalist development in Europe. Europe enjoyed no innate cultural or institutional supremacy—it just happened to be in the right place at the right time.

Likewise, Pomeranz emphasizes two points of good fortune that enabled the rise of the West. First, Britain's industrial revolution was enabled by the transmission of land-saving products from the American colonies, without which Britain would have been unable to industrialize (since more domestic resources would have been diverted away from industrialization back into agriculture).[25] And second, the "great divergence" between China and Britain was a product of good fortune once again. For while China's coal mines were shallow and arid, Britain's were deep and flooded.[26] Moreover Britain's mines, unlike China's, were conveniently located close to the main iron production and market centres. Above all, Britain's mines being deep and flooded necessitated the invention of the British steam engine to pump the water out. And the rest, as they say, is (modern industrial Western) history.[27] For Pomeranz, the British *adapted* to the contingent ecological problem of deep and flooded mines by inventing the steam engine, no less than they were able to adapt to the ecological constraints of industrialization by importing land-saving products from the Americas. A third approach can be discerned in the work of Jack Goldstone.[28] He, in fact, grants Europe some cultural agency while retaining a perspective that emphasizes contingent "freakish accident." Goldstone claims that the mid-seventeenth century is the moment when the origins of the later divergence between the East and the West can be discerned. In responding to common problems of internal rebellion, the Ottoman Empire and China restored traditional, inflexible cultural norms that subsequently held them back. By contrast, a "freakish accident" occurred in England. This unexpected twist saw the invasion of England by the Protestant, William of Orange, which subsequently prevented a domination of Europe by Catholicism. No less importantly, the nature of how Protestantism unfolded in England meant that monarchs could not monopolize political power, the net result of which was the creation of an open and tolerant cultural milieu within which the Scientific Revolution could flourish. This was given further impetus by the Anglican Church's nurturing of the new mechanical worldview. These two developments provided a permissive environment for the rapid progress of technological inventiveness, which in turn led on to invention of the steam engine to solve the problem of deep and flooded mines. And, as with Pomeranz, the rest was (modern industrial Western) history.

Thus for Goldstone, the British *adapted* to the mid-seventeenth-century crisis by replacing the Catholic King James II by the Protestant, William of Orange. The upshot of this created, albeit through a series of chance factors, a permissive environment in which science flourished. And he matches the adaptive/contingency argument of Pomeranz concerning the invention of the steam engine to solve the problem of flooded mines. Another way of putting this is to say that adapting to various crises and challenges had the largely unintended consequence of leading to the industrial breakthrough. Certainly this knocks out the Eurocentric claim that the breakthrough was the *inevitable culmination* of centuries of exceptional institutional and cultural/moral progress. By contrast, neo-Eurocentrics view the scientific breakthrough as a natural function of superior and innate cultural ideas that enabled Europe to borrow and learn from others in an open and tolerant milieu, thereby promoting European development from the thirteenth century (Duchesne) or from ancient Greek times (Warraq).

This means that non-Eurocentrism *can* offer an account of European adaptiveness. But, as I shall now explain, non-Eurocentrism is lacking for the most part a *theory* and *explanation* of European adaptiveness; a lacuna that strikes a hole in the body of this literature. How so? There is little doubting the point that most non-Eurocentric scholars are highly skeptical of any argument that privileges *internal* European attributes, precisely because they see in this an unacceptable expression of Western exceptionalism. Typical here are the words of James Blaut:

> Europe's environment is *not* better than the environment of other places—
> not more fruitful, more comfortable, more suitable for communication and
> trade, and the rest. Europe's culture did *not*, historically, have superior

traits, traits that would lead to more rapid progress than that achieved by
other societies: individual traits like inventiveness, innovativeness, ambi-
tiousness, ethical behavior, etc.; collective traits like the family, the market,
the city. The rise of Europe cannot be explained in this Eurocentric way.[29]

Or as Frank expressed it, "Europe did not pull itself up by its own economic bootstraps, and certainly not thanks to any kind of European
'exceptionalism' of rationality, institutions, entrepreneurship, technology, geniality, in a word—of race."[30] In essence, writers like Blaut or
Frank reject the notion of a Europe whose breakthrough to modernity is
self-generated. But in fairness one is, I think, indeed entitled to ask: surely
Europe did something "right," for was it not the Europeans rather than
the Easterners who made the breakthrough?

The same claim can be made with respect to Pomeranz's *Great
Divergence*. The problem here is that the adaptability argument is embedded within a functionalist narrative. Pomeranz accounts for the divergence by arguing ultimately that Britain's mines were deep and flooded
and "functionally required" the invention of the steam engine.[31] But this
does not *explain* what it was that enabled its invention. This is rendered
yet more problematic by the fact that many of China's mines were in
fact not just deep but were flooded.[32] Invoking "functional-ecological
necessity" or pure ecological contingency does not explain the factors
that enabled the Europeans rather than the Chinese to invent the steam
engine in the first place. For "necessity may indeed be the mother of
invention, but it is well to remember that 'it is ideas which make necessity conscious'."[33] Likewise Frank emphasizes the point that Britain
invented the steam engine and other "labour-saving" technologies out
of the structural-functional imperatives imposed by the global economic
downturn of the 1760s (as was noted above) but tells us nothing about
what enabled this. The general problem here is that structural economic-
or ecological-functionalism explains little but provides only circular logic
or tautology.

Thus it seems to me that the elephant in the non-Eurocentric room
takes the form of advocating functional necessity in place of an explanation that has theoretical bases. I would argue that dialogical globalism
also suffers in this respect. For the thrust of it is to point up the manifold
Eastern inventions that diffused across to the West to eventually enable
its rise into modernity. But there is no explanation of what the Europeans
did to adapt these resources to higher ends, which in turn implies at
the very least some kind of inventive capacity. It is merely assumed that
they developed this capacity. And this in turn raises another thorny issue
that needs to be explained. That is, if the East did so much to generate
the resource portfolios that enabled the creation of modernity in the
West, why then did the East not go on to make the breakthrough itself?

Non-Eurocentric thinkers have engaged in various intellectual acrobatics to avoid answering this question. And the reason for this is that they are loathe to attribute the necessary adaptive/inventive properties to the West on account of the point that this seems to take us back into the so-called cul-de-sac of Eurocentrism and a celebration of all things Western. So what then is to be done?

WHAT IS TO BE DONE?

The first thing to note here is that Jack Goldstone has already had a stab at doing this. For him, as noted above, it was the tolerant and open institutional milieu within which science could flourish that provided the key to Britain's breakthrough. Moreover, he also counterfactually claims that such an open institutional environment was lacking in the major Eastern regions. To some non-Eurocentrics, however, this will seem like a step back into Eurocentrism, and it certainly echoes the neo-Eurocentric arguments and approach. But, if so, then this raises the following profound conundrum for non-Eurocentrism: is it possible to produce a non-Eurocentric theoretical explanation of the things that "Europe did right," without falling back into the trap of Eurocentrism and neo-Eurocentrism? Put more specifically, we need to know what the Europeans did right to put all the Eastern resource portfolios together to eventually make the breakthrough on the one hand, as well as answering why they sought to achieve all this on the other. And simultaneously, we need to confront the other elephant in this (already overcrowded) room, namely: why did the Easterners not press on into modernity? Of course, non-Eurocentrics will be searching for answers that avoid endogenous *institutional blockages* in the case of the "fall of the East" or *institutional advantages* in the case of the West. Another way of putting this is to say that non-Eurocentrism, and especially global dialogism, has in effect eradicated European agency in its quest to reinstate Eastern agency. But failure to reinstate some degree of European agency means that we end up with an Occidentalist narrative that is problematic not least because it cannot offer a theory or explanation for the rise of the West as much as the "fall of the East."

My own first-cut solution to this challenge is one that only half answers these questions and it is to the missing part that I am looking to work upon. My argument would be that the West made it to the top because it developed a kind of "Eurocentric restlessness" that was in turn founded upon a certain insecurity vis-à-vis other civilizations. This argument provides the background for the West's energetic rise and reinstates some agency to the West. It helps explain and address the "why question" but it does *not* explain the "how question"—that is, *how* did the Europeans make it to the top and what did the Europeans do to put all the Eastern resource portfolios to higher ends? So there is a prospective answer to the "why" but not the "how" question. Nevertheless, here I want to focus on the "why question." This is significant because it entails focusing on the lost issue of European agency. I undertake this by considering the role of identity-formation; something that is also absent from non-Eurocentric analysis. Indeed, in contrast to the materialism of Blaut and Pomeranz, Frank and Abu-Lughod, I accord a certain autonomy not only to ideas and culture—as does Goldstone—but also to identity. This emphasis on identity is certainly charted in considerable detail in my 2004 book, *The Eastern Origins of Western Civilisation*, but it was done in the context of specifically explaining Europe's drive to imperialism after 1492. Here I want to extend it further to explain the causes of Europe's adaptive capacity to assimilate Eastern resource portfolios and to eventually deploy them to "higher economic ends" to promote the industrial breakthrough.

The argument here begins principally in the year 1453, when the perception of an Ottoman "identity threat" confronted Christendom, leading the Papacy to issue a series of papal bulls. These effectively ordered the Iberians to undertake another round of Crusades by expanding outward to make links with the Catholic priest-king, Prester John, in the "Indies" with whom they were to forge a Christian alliance and attack the Muslims from the rear. Columbus, of course, dutifully sailed westward in search of the (East) Indies.[34] But in literally "stumbling" across the Americas, the tragic encounter with the Native Americans led the Europeans to believe—for the first time in a millennium—that they were superior to another civilization. And this emergent sense of superiority was exacerbated through their subsequent exploitation of Black African labor. The sixteenth century in European history saw not the rise of Europe to the top, since Europe was still considerably behind China, India, Islamic West Asia, and an emergent Japan under the Tokugawa (after 1600), but the rapid rise of a new aggressive European identity. This identity furnished the Europeans with a more urgent drive to "catch up" with the still vastly more advanced Asian world. This was reinforced as first the Portuguese, then the Dutch, and finally the English arrived in the Indian Ocean "all-conquering," only to be delivered a rude shock. For they quickly found that they had no choice but to cooperate with the more advanced Islamic/Asian merchants and Asian rulers right down to 1800 just to gain a meager slice of the Indian Ocean trade.[35] Overall, with the exception of the Americas and sub-Saharan Africa, the Europeans endured an inferiority complex vis-à-vis the Muslims, Indians, and Chinese. And overcoming this inferiority complex led to an aggressive mentality—a Eurocentric restlessness—that led on to the pursuit of development at any cost.

Crucial to my argument is that after the mid-fifteenth century Europe was a "late-developer" civilization.[36] And late-developers enjoy the many "advantages of backwardness."[37] This refers to the fortuitous situation whereby late-developers benefit from the advanced technologies that were pioneered previously by the early developers. The key early developers of central significance here were the Muslims, North Africans, Indians, and Chinese.[38] Thus the Europeans, having begun to access Islamic scientific ideas around 1085, resorted to working with them more urgently, which in turn helped promote the Renaissance and later on the Scientific Revolution.[39] This coincided with Europe's development of military power during its Military Revolution between 1550 and 1660, which centered on the deployment of the gun, gunpowder, and cannon. But these were in fact invented during China's Military Revolution (c. 850–c.1290), and were subsequently assimilated and then later adapted to higher ends as the Europeans became intent on conquering other parts of the world.

Then, from the late-seventeenth century onward the Europeans began to voraciously consume and appropriate Chinese ideas and technologies, which in turn directly promoted the British agricultural and industrial revolutions. It is certainly the case that Chinese ideas had been relayed back to Europe mainly via the friars who visited China in the thirteenth century (with Marco Polo being the most famous). But it took a long while for these ideas to have much impact partly because Europe had not yet developed the aggressive mentality to use them, and partly because Marco's reports appeared incredible to the then backward "red-haired barbarians." By the eighteenth century, however, Chinese ideas such as rationality and laissez faire (the French translation of *wu-wei*), which had been around for almost 2,000 years in China, directly entered the vocabulary of the Enlightenment *philosophes*—though at least in this instance many of these *philosophes*, such as Voltaire and Quesnay, gave full credit to the Chinese. The Europeans fed off the many pamphlets and books that flooded Europe in the seventeenth century, which in aggregate revealed the manifold technologies and advanced features of Chinese civilization. These included seed-drills and horse-hoeing husbandry, crop rotation systems, ideas that enabled the invention of the steam engine, canal pound locks, bulkheads/watertight ship compartments, steel production techniques, and many more. And not infrequently, the actual technology itself was brought over, as in the curved iron moldboard plough and rotary winnowing machine. Moreover, many Europeans traveled to China specifically to learn of their inventions and industrial/agricultural production processes. These multiple transmission circuits provided a crucial knowledge base that helped promote the British agricultural and industrial revolutions.[40]

Marshall Hodgson once noted in passing that the Occident was "the unconscious heir of the industrial revolution of Sung China."[41] As should be apparent by now, but for the word "unconscious" I concur, given that the British *consciously* acquired Chinese ideas and technologies to eventually adapt them to higher industrial ends. Perhaps, then, the British could be likened here to the Eurocentric characterization of the Japanese after 1945: that they developed a highly adaptive capacity and were excellent at copying and refining further others' ideas and technologies. For in the end the Europeans adapted some of these resource portfolios to higher ends. This is what one expects from an adaptive late-developer civilization, though Europe was no ordinary example of one. For ultimately, this was fueled by a relentless Eurocentric restlessness.

As I argued in detail in my 2004 book, this internal adaptive drive found its external expression in the drive to imperialism as a rapidly rising racist identity fueled the exploitation of Eastern resources—land, labour, raw materials, land-saving products, bullion, and markets—all of which were vital to supplying the British industrial revolution.[42] This was the second major input that the "non-West" provided in enabling British industrialization. Moreover, in the British case, this was supported by a highly fiscal-militarist interventionist late-developer state that applied heavy protectionism at home and tolerated protectionism and wars in continental Europe, while imposing free trade, backed up by naval cannon, on the imperial economies abroad.[43] All in all, the internal and external adaptive drives can be seen to underpin the all-important British industrial breakthrough, which in turn was underpinned by this Eurocentric restlessness.

But this in turn raises the counterfactual question based on the issue of the "great divergence": can this identity-based framework be deployed to explain why China did not go on and make the breakthrough? If China lacked anything it was an equivalent identity-based drive to that of Europe's at this crucial historical moment. China was reluctant to engage in imperialism, since its identity was a defensive construct that was designed to maintain the emperor's legitimacy in the eyes of his/her domestic population.[44] This was internationalized, as was Europe's identity, though it led not to imperialism but to the Chinese tribute system. For its social function required gaining the *allegiance* of the "barbarian world" so that the emperor could maintain the legitimacy of the Chinese state and society.[45] This means that the Messianistic drive that spurred on European development, internally and externally, was indeed lacking in China. But we should not write off China's capacity in the post-Sung period either to innovate,[46] or to learn from foreign ideas.[47] And nor should this argument be conflated with the standard claim,[48] that China withdrew from the world trading system into its own regressive Sinocentric closure after the proclamation of the official ban in 1434 (as

I charted in my 2004 book).[49] There was, however, one sense in which the Chinese broadly "withdrew": they withdrew not from global trade but *abstained* from the imperial power politics that would shortly grip the Europeans after 1492.

All in all, I would agree with Peer Vries's *rejection* of the familiar Eurocentric conclusion that Europe's "ethic of world mastery" contrasted with China's passive ethic of "adjustment to the world," as Max Weber originally declared.[50] This argument also serves to problematize Ibn Warraq's emphasis on Europe's tolerance and cosmopolitan multiculturalism as underpinning Europe's rise. For Europe's catch-up in the 1492–1850 period directly coincided with the rise of a new Eurocentric European identity. And at the risk of sounding flippant, concerning the "why question" I would conclude that the Europeans did indeed do something "right'—they constructed a late-developer Eurocentric identity that fueled a relentless and ruthless adaptive drive to appropriate and plunder the many resources and resource portfolios of the Eastern societies to achieve modernization. This, I would venture as a first step toward developing a theoretical account and explanation of the rise of the West, though I fully concede that without an answer to the question as to *how* the Europeans assimilated others' resource portfolios to higher ends, we have not succeeded in rescuing the non-Eurocentric enterprise.

I have also used an identity-based argument to venture a possible reason as to why China did not make the breakthrough. Even so, we need to inquire further into Chinese science not least to consider whether it could have supported a modernization drive in general, and whether the Chinese could have invented the steam engine had the British not got there first (though this is not to say that the invention of the steam engine is the ultimate factor in the rise of the West). However, thought needs to be extended to considering other Eastern civilizations here. India, for example, was undoubtedly blighted by British colonialism, though such a logic would not fit well with China or West Asia. But it still begs the question: were China, West Asia, and India capable of making the breakthrough themselves? And how are we to deal with the point that the European Renaissance and scientific revolutions were significantly shaped by Islamic ideas, only to find that this initial West Asian thrust was curtailed some time after the sixteenth century, perhaps at the behest of Islamic religious authorities as Eurocentrism/neo-Eurocentrism claims?

Either way, though, I hope that in at least raising these issues, we might begin to think about possible non-Eurocentric answers and push global dialogism out of its current impasse. And given my hunch that European science had something to do with the rise of the West—as Goldstone argues—this might constitute an excellent starting point to launch this new voyage.

END NOTES

1. Edward W. Said, *Power, Politics, and Culture* (London: Bloomsbury, 2004), 260–261, 424.
2. Ernst Gombrich, "Eastern Inventions and Western Response," *Daedalus* 127, no. 1 (1998): 202.
3. Ibn Warraq, *Defending the West* (New York: Prometheus Books, 2007), 134.
4. S. P. Huntington, *The Clash of Civilizations and the Remaking of World Order* (New York: Touchstone, 1996).
5. P. T. Jackson, "'Civilization' on Trial," *Millennium* 28, no. 1 (1999): 141–153.
6. Huntington, *Clash of Civilizations*, 43–44.
7. Jessica Benjamin, *The Bonds of Love* (New York: Pantheon, 1988).
8. Karl Pearson, *National Life from the Standpoint of Science* (London: Adam & Charles Black, 1905); Lothrop Stoddard, *The Rising Tide of Color against White World Supremacy* (New York: Charles Scribner's Sons, 1920). For a full discussion of this see John M. Hobson, *The Eurocentric Conception of World Politics: Western International Theory 1760–2010* (Cambridge: Cambridge University Press, 2012), chs. 4, 5, 6–7.
9. For a general discussion see, for example, J. M. Hobson, *The Eastern Origins of Western Civilisation* (Cambridge: Cambridge University Press, 2004); J. M. Hobson, "Deconstructing the Eurocentric Clash of Civilizations: De-Westernizing the West by Acknowledging the Dialogue of Civilizations," in *Civilizational Identity*, edited by Martin Hall and P. T. Jackson, 149–165 (New York: Palgrave Macmillan, 2007); Jack Goody, *Islam in Europe* (Cambridge: Polity Press, 2004). For a discussion of this in relation to science and mathematics respectively see: Arun Bala, *The Dialogue of Civilizations in the Birth of Modern Science* (Houndmills: Palgrave Macmillan, 2006); George G. Joseph, *The Crest of the Peacock* (London: Penguin Books, 1992).
10. For a full discussion see Hobson, "Deconstructing the Eurocentric Clash of Civilizations"; Hobson, *Eurocentric Conception of World Politics*, ch. 11.
11. For example, Janet L. Abu-Lughod, *Before European Hegemony* (Oxford: Oxford University Press, 1989); James M. Blaut, *The Colonizer's Model of the World* (London: Guilford Press, 1993); Jack Goody, *The East in the West* (Cambridge: Cambridge University Press, 1996); Andre Gunder Frank, *ReOrient* (Berkeley, CA: University of California Press, 1998); Kenneth Pomeranz, *The Great Divergence* (Princeton, NJ: Princeton University Press, 2000); Jack Goldstone, *Why Europe? The Rise of the West in World History, 1500–1800* (New York: McGraw Hill Higher Education, 2009).
12. Ricardo Duchesne, "Asia First?," *The Journal of the Historical Society* 6, no. 1 (2006): 69–91; J. M. Hobson, "Explaining the Rise of the West: A Reply to Ricardo Duchesne," *The Journal of the Historical Society* 6, no. 4 (2006): 579–599.
13. Duchesne, "Asia First?"; Warraq, *Defending the West*; Toby E. Huff, *The Rise of Early Modern Science: Islam, China and the West* (Cambridge: Cambridge University Press, 2003). Note, however, that in his most recent book Huff seems to have retreated from a *neo*-Eurocentric approach into a more monolithic and hardened Eurocentrism; Toby E. Huff, *Intellectual*

Curiosity and the Scientific Revolution (Cambridge: Cambridge University Press, 2011).

14. Duchesne, "Asia First?," 76, 79.
15. Carlo M. Cipolla, *Before the Industrial Revolution* (New York: W. W. Norton, 1976).
16. Francis Oakley, *The Crucial Centuries* (London: Terra Nova Editions, 1979), 100.
17. Huff, *Rise of Early Modern Science*, 13.
18. Cipolla, *Before the Industrial Revolution*, 180.
19. Warraq, *Defending the West*, 55–296.
20. Warraq, *Defending the West*, 62.
21. Walter Burkert, *Babylon, Memphis, Persepolis: Eastern Contexts of Greek Culture* (Cambridge, MA: Harvard University Press, 2004), 12, cited in Warraq, *Defending the West*, 71.
22. Frank, *ReOrient*.
23. Abu-Lughod, *Before European Hegemony*, 361.
24. Blaut, *Colonizer's Model*, 180–183.
25. Pomeranz, *Great Divergence*, ch. 6.
26. Ibid., 65.
27. Ibid., esp. ch. 1.
28. Jack A. Goldstone, "The Rise of the West—or Not? A Revision to Socioeconomic History," *Sociological Theory* 18, no. 2 (2000): 175–194.
29. James M. Blaut, *Eight Eurocentric Historians* (London: Guilford Press, 2000), 1–2, emphases in the original.
30. Frank, *ReOrient*, 4, and esp. ch. 7.
31. Pomeranz, *Great Divergence*, 65.
32. Peter J. Golas, *Science and Civilization in China*, vol. 5, no. 13 (Cambridge: Cambridge University Press, 1999), 186, 336; Hobson, *Eastern Origins*, 207–208; Peer Vries, "Is California the Measure of All Things? A Rejoinder to Ricardo Duchesne, 'Peer Vries, the Great Divergence, and the California School: Who's In and Who's Out?'," *World History Connected* 2, no. 2 (2005), n. 50.
33. Oakley, *Crucial Centuries*, 99–100.
34. Hobson, *Eastern Origins*, 135–137, 162–168.
35. Hobson, *Eastern Origins*, ch. 7.
36. Note that Frank too deploys this term but in a somewhat different context. Moreover, he dates the European late-developer strategy to the nineteenth century; *ReOrient*, esp. 318–319.
37. Alexander Gerschenkron, *Economic Backwardness in Historical Perspective* (Cambridge, MA: Harvard University Press, 1962).
38. Hobson, *Eastern Origins*, chs. 2–4; cf. E. L. Jones, *Growth Recurring* (Oxford: Clarendon, 1988).
39. Hobson, *Eastern Origins*, 173–183; Goody, *Islam in Europe*, 56–83; Bala, *Dialogue of Civilizations*; S. M. Ghazanfar, *Islamic Civilization* (Lanham, MD: Scarecrow Press, 2006).
40. Hobson, *Eastern Origins*, ch. 9.
41. Marshall G. S. Hodgson, *The Venture of Islam, III* (Chicago, IL: University of Chicago Press, 1974), 197.

42. Hobson, *Eastern Origins*, chs. 10–11.
43. Cf. Ha-Joon Chang, *Kicking Away the Ladder* (London: Anthem Press, 2002).
44. Hobson, *Eastern Origins*, esp. 68–70, 307–308.
45. Y. Zhang, "System, Empire and State in Chinese International Relations," in *Empires, Systems and States*, edited by M. Cox, K. Booth, and T. Dunne (Cambridge: Cambridge University Press, 2001).
46. For example, Pomeranz, *Great Divergence*, 47–48; Jones, *Growth Recurring*, 143–144.
47. Joanna Waley-Cohen, *The Sextants of Beijing*. New York: W. W. Norton, 1999.
48. As does David Landes, *The Wealth and Poverty of Nations* (London: Little, Brown and Company, 1998), 96.
49. Hobson, *Eastern Origins*, ch. 3.
50. Peer Vries, *Via Peking Back to Manchester* (Leiden: CNWS, 2003), 35.

BIBLIOGRAPHY

Abu-Lughod, Janet L. 1989. *Before European Hegemony*. Oxford: Oxford University Press.
Bala, Arun. 2006. *The Dialogue of Civilizations in the Birth of Modern Science*. Basingstoke: Palgrave Macmillan.
Benjamin, Jessica. 1988. *The Bonds of Love*. New York: Pantheon Books.
Blaut, James M. 1993. *The Colonizer's Model of the World*. London: Guilford Press.
———. 2000. *Eight Eurocentric Historians*. London: Guilford Press.
Burkert, Walter 2004. *Babylon, Memphis, Persepolis: Eastern Contexts of Greek Culture*. Cambridge, MA: Harvard University Press, 2004, p. 12.
Chang, Ha-Joon. 2002. *Kicking Away the Ladder*. London: Anthem Press.
Cipolla, Carlo M. 1976. *Before the Industrial Revolution*. New York: W. W. Norton.
Duchesne, Ricardo. 2006. "Asia First?." *The Journal of the Historical Society* 6, no. 1.
Frank, Andre Gunder. 1998. *ReOrient*. Berkeley, CA: University of California Press.
Gerschenkron, Alexander. 1962. *Economic Backwardness in Historical Perspective*. Cambridge, MA: Harvard University Press.
Ghazanfar, S. M. 2006. *Islamic Civilization*. Lanham, MD: Scarecrow Press.
Golas, Peter J. *Science and Civilization in China*, vol. 5 part 13. Cambridge: Cambridge University Press, 1999.
Goldstone, Jack A. 2000. "The Rise of the West—or Not? A Revision to Socioeconomic History." *Sociological Theory* 18, no. 2.
———. 2009. *Why Europe? The Rise of the West in World History, 1500–1800*. New York: McGraw Hill Higher Education.
Gombrich, Ernst. 1998. "Eastern Inventions and Western Response." *Daedalus* 127, no. 1.
Goody, Jack. 1996. *The East in the West*. Cambridge: Cambridge University Press.

————. 2004. *Islam in Europe*. Cambridge: Polity Press.

Hobson, J. M. 2004. *The Eastern Origins of Western Civilisation*. Cambridge: Cambridge University Press.

————. 2006. "Explaining the Rise of the West: A Reply to Ricardo Duchesne." *The Journal of the Historical Society* 6, no. 4.

————. 2007. "Deconstructing the Eurocentric Clash of Civilizations: De-Westernizing the West by Acknowledging the Dialogue of Civilizations." In *Civilizational Identity*, edited by Martin Hall and P. T. Jackson. New York: Palgrave Macmillan.

————. 2012. *The Eurocentric Conception of World Politics: Western International Theory 1760–2010*, Cambridge: Cambridge University Press.

Hodgson, Marshall G. S. 1974. *The Venture of Islam, III*. Chicago, IL: University of Chicago Press.

Huff, T. E. 2003. *The Rise of Early Modern Science: Islam, China and the West*. Cambridge: Cambridge University Press.

————. 2011. *Intellectual Curiosity and the Scientific Revolution*. Cambridge: Cambridge University Press.

Huntington, S. P. 1996. *The Clash of Civilizations and the Remaking of World Order*. New York: Touchstone.

Jackson, P. T. 1999. "'Civilization' on Trial" *Millennium* 28, no. 1.

Jones, E. L. 1988. *Growth Recurring*. Oxford: Clarendon.

Joseph, George G. 1992. *The Crest of the Peacock*. London: Penguin.

Landes, David. 1998. *The Wealth and Poverty of Nations*. London: Little, Brown and Company.

Oakley, Francis. 1979. *The Crucial Centuries*. London: Terra Nova Editions.

Pearson, Karl. 1905. *National Life from the Standpoint of Science*. London: Adam & Charles Black.

Pomeranz, Kenneth. 2000. *The Great Divergence*. Princeton, NJ: Princeton University Press.

Said, Edward W. 2004. *Power, Politics, and Culture*. London: Bloomsbury.

Stoddard, L. 1920. *The Rising Tide of Color against White World Supremacy*. New York: Charles Scribner's Sons.

Vries, Peer. 2003. *Via Peking Back to Manchester*. Leiden: CNWS.

————. 2005. "Is California the Measure of All Things Global? A Rejoinder to Ricardo Duchesne, 'Peer Vries, the Great Divergence, and the California School: Who's In and Who's Out?'." *World History Connected* 2, no.2.

Waley-Cohen, Joanna. 1999. *The Sextants of Beijing*. New York: W. W. Norton.

Warraq, Ibn. 2007. *Defending the West*. New York: Prometheus Books.

Zhang, Y. 2001. "System, Empire and State in Chinese International Relations." In *Empires, Systems and States*, edited by M. Cox, K. Booth, and T. Dunne. Cambridge: Cambridge University Press.

2

A PASSAGE TO INFINITY: THE CONTRIBUTION OF KERALA TO MODERN MATHEMATICS*

George Gheverghese Joseph

INTRODUCTION

Two powerful tools contributed to the creation of modern mathematics in the seventeenth century: the discovery of the general algorithms of calculus and the development and application of infinite series techniques. These two streams of discovery reinforced each other in their simultaneous development with each serving to extend the range and application of the other. According to existing literature, the calculus that resulted were invented independently by Newton and Leibniz, building on the works of their European predecessors such as Fermat, Taylor, Gregory, Pascal, and Bernoulli during the preceding half-century.[1] But what appears to be less well-known is that certain fundamental elements of this calculus including numerical integration methods and infinite series derivations for π and of certain trigonometric functions such as sin x, cos x, and tan^{-1} x (the so-called Gregory series) were already known about 250 years earlier in Kerala, South India. In recent years there has been some acknowledgment of this fact. But such acknowledgments are quite rare.[2]

There are several questions worth exploring about the Kerala work, apart from its technical content. This chapter begins with a brief survey of the background to this work and the mathematical motivation behind the interest in a particular series (namely, the arctan series and its special case—the π series). The second part of this chapter examines the epistemology of the "calculus" of the Kerala School and its possible transmission to Europe. In general terms, we examine the hypothesis that the arrival of Vasco da Gama in Calicut (in present-day Kerala) at the end of the fifteenth century not only short-circuited the traditional

Arab route for spices, but also the traditional Arab route into Europe of Indian mathematics and astronomy.

THE BACKGROUND

The direct inspiration for Kerala mathematics is found in the works of Aryabhata and his commentators, notably Bhaskara I. In 499 CE at the age of 23, Aryabhata composed his seminal text *Aryabhatiya*. The influence of the astronomical and mathematical ideas in this text, both inside and outside India, cannot be overestimated. This was particularly so with respect to the work of the Kerala School, founded by Madhava (fl. 1340–1425) and developed by his disciples, Nilakantha Somayaji (fl. 1444–1500), the author of *Tantrasangraha* ("A Digest of Scientific Knowledge"); Sankara Variyar (1500–1560) and Narayana (c. 1500–1575), the authors of *Kriyakramakari* ("Operational Techniques"); Jyesthadeva (fl. 1500–1610), the author of the *Yuktibhasa* ("An Exposition of the Rationale"); Acyuta Pisharoti (fl. 1550–1621), the author of the *Sphutanirnaya*; Putumana Somayajin (fl. 1660–1740), the author of *Karanapaddati* ("A Manual of Performances in the Right Order"); and others. The last recognized text of the Kerala School that added little to previous work was *Sadratnamala* ("A Garland of Pearls") of Sankara Varman (1800–1838). These authors formed part of a tradition of continuing scholarship in Kerala over a period of four hundred years from the birth of Madhava around 1340 to the probable death of Putumana Somayajin in 1740.

Madhava has been generally acknowledged by those who came after him, to be responsible for key developments in infinite series expansions. These developments included the "Gregory-Taylor" series for sine, cosine, and arctan functions, with accurate remainder terms, and a numerically efficient algorithm, leading to accurate tables for sines and cosines.[3] The Kerala work also encompassed other innovations in mathematics and astronomy, including obtaining the derivatives of various trigonometric functions in the *Tantrasangraha* and in the *Sphutanirnaya* to compute the instantaneous velocities of the Sun, the Moon, and other planets, as well as the implicit value of π correct to 9, 10, (and later 17) decimal places.[4]

Information on the lives of the members of the Kerala School is fragmentary at present. Madhava and his disciples probably lived and worked in large compounds called *illams* situated mainly in Brahmin settlements, concentrated into an area less than five-hundred square miles north of Cochin. Many of these *illams* belonged to households owning large landed properties. While these estates were farmed by workers or tenants from lower castes, the Nambuthris (a high-ranking subcaste of Brahmins), and particularly the younger sons, enjoyed considerable leisure and passed their time in study and ritual observances.[5]

These *illams* provided a base for the education of the young in Sanskrit works, including mathematical and astronomical classics (notably the *Aryabhatiya* of Aryabhata and its commentaries). Not only was the transmission of knowledge accomplished in these *illams*, but they also provided a center for research and scholarship. Sometimes the scholars wrote commentaries on the classics and in those commentaries they appended their own discoveries as additions and supplements.[6] The close proximity between the *illams* and their relationship with temples during a long period of political stability in the region created a favorable climate for continuing intellectual development based on generations of teacher-student relationships.[7]

To obtain a flavor of the mathematics of the Kerala School, consider the following quotation from Jyesthadeva's *Yuktibhasa* relating to the arctan series. Note that capital Sine (Cosine), sometimes called Indian sine (cosine), is the product of radius and modern sine (cosine).[8]

> The product of given Sine and the radius divided by the Cosine is the first result. From the first (and then the second, third,...etc.) results, obtain (successively) a sequence of results by taking the square of the Sine as the multiplier and the square of the Cosine as the divisor. Divide (the above results) in order by the odd numbers 1, 3, 5...etc. to get the (full sequence of) terms. From the sum of the odd terms, subtract the sum of the even terms. (The result) becomes the arc. In this connection, it is laid down that the (Sine) of the arc of (that of) its complement, whichever is smaller, should be taken here (as the "given Sine"); other wise, the terms, obtained by the (above) repeated process will not tend to a vanishing magnitude.

Translated into modern symbolic notation, this reads:

$$\theta = \tan\theta - (1/3)\tan^3\theta + (1/5)\tan^5\theta - \ldots$$

There are a couple of interesting features about this passage from the *Yuktibhasa*. First, its most reliable version, unusually in the Indian context, is in Malayalam (the local language of Kerala) and not in Sanskrit.[9] Second, the series whose construction is explained by the quotation is now known as the Gregory series named after James Gregory, a Scottish mathematician, who examined it in 1671. It may more accurately be called the Madhava-Gregory series, since it has been attributed by a number of members of the Kerala School to their founder and easily precedes the work of Gregory by about three hundred years.

There is a further aspect of the *Yuktibhasa* that is important. As the very name *Yuktibhasa* implies, unlike many other Indian mathematical works of that and earlier periods, the text contains a detailed exposition of the rationale (or proofs) expressed in a verbal form, consisting of a mixture of technical terms and *katapayadi* notation, the latter being a refinement of Aryabhata's alphabet-numeral system of notation.[10] It is from this and other texts, notably the *Kriyakramakari*, that one can put together the derivation of the arctan series according to Kerala mathematicians. A detailed derivation is not attempted here. They have already been covered elsewhere in the literature, notably in the writings of Rajagopal and his collaborators, mentioned earlier and listed in the bibliography at the end of this chapter.[11]

MOTIVATION AND METHOD

A primary mathematical motivation for the Kerala work on infinite series arose from a recognition of the impossibility of arriving at an exact value for the circumference of a circle given the diameter. Nilakantha explained in his *Aryabhatiyabhasya* (a commentary on Arybhata's *Aryabhatiya*) why only an approximate value of the circumference could be obtained:

> If the diameter can be measured without a remainder, the circumference measured by the same unit (of measurement) will leave a remainder. Similarly, the unit which measures the circumference without a remainder will leave a remainder when used for measuring the diameter. Hence, the two measured by the same unit will never be without a remainder. Though we try very hard we can reduce the remainder to a small quantity but never achieve the state of "remainderlessness." This is the problem.

This explanation was prompted by a passage in *Aryabhatiya*. Verse 10 of the section on *Ganita* states:

> Add 4 to 100, multiply by 8, and add 62,000. The result is *approximately* the circumference of a circle whose diameter is 20,000.

It was the word "approximately"[12] that gave food for thought. And the strategy recommended in the *Kriyakramakari* of Sankara and Narayana:

> Thus even by computing the results progressively, it is impossible theoretically to come to a final value. So, one has to stop computation at that stage of accuracy that one wants and take the final result arrived at ignoring the previous results.

The approach involved finding the length of an arc by approximating it to a straight line. Known today as the method of direct rectification of an arc of a circle, it involves summation of very small arc segments and reducing the resulting sum to an integral. The tangent is divided up into equal segments while at the same time forcing a subdivision of the arc into *unequal* parts. This is required, since the method involves the summation of a large number of very small arc segments, traditionally achieved in European and Arab mathematics by the "method of exhaustion," where there was a subdivision of an arc into *equal* parts. The adoption of this "infinite series" technique rather than the "method of exhaustion" was not due to ignorance of the latter in Kerala mathematics. But, as Jysthadeva implies in the *Yuktibhasa*, the former avoids tedious and time-consuming root-extractions.[13]

There are other interesting aspects regarding the "tool kit" used by the Kerala mathematicians. The derivation of the arctan series employs two results in elementary mathematics that have a long history in India: (1) The right-angle theorem, known in the West as the Pythagorean theorem, the knowledge of which dates back to the *Sulbasutras* (variously dated 500–800 BC); and (2) the properties of similar triangles that is little more than a geometrical version of the "rule of three" *(trairasika)* of which probably the first systematic treatment is found in the *Bakhshali* manuscript (variously dated from AD 200–700).[14]

The Kerala derivation deploys an ingenious iterative resubstitution procedure[15] to obtain the binomial expansion for the expression $1/(1+x)$ and then proceeding through a number of repeated summations (*varamsamkalithas*) of series, arrives at what must be the most remarkable part of the derivation, an intuitive leap as it were leading to the asymptotic formula, expressed in modern notation as follows:[16]

$$\lim_{n \to \infty} \frac{1}{n^{k+1}} \sum_{i=1}^{n} i^k = \frac{1}{k+1}, \quad k = 1, 2, 3, \ldots \tag{1}$$

It was soon realized that the series

$$\pi/4 = 1 - 1/3 + 1/5 - \ldots \tag{2}$$

was not useful in obtaining accurate estimates of the circumference for specified diameters (i.e., in estimating π) because of the slowness of the convergence of the series. This gave impetus to developments in two directions: (1) obtaining rational approximations by applying corrections to partial sums of the series; and (2) obtaining more rapidly converging series by transforming the original series. There was considerable work in both directions.[17] What this work exhibits is a measure of understanding of the concept of "convergence," of the notion of rapidity of convergence and a clear awareness that convergence can be speeded up by transformations.

As an illustration of the remarkable efficiency of some of the corrections suggested, consider the following example from *Yuktibhasa*. What is required is to evaluate the circumference of a circle with a diameter of 10^{11}. *Without* the correction, and taking 19 terms on the right-hand side of equation (2), the implicit value for the circumference as the ratio of the diameter is around 3.194. However, incorporating one of the corrections[18] gives the circumference as $3.1415926529 \times 10^{11}$ that is correct to 8 places. A search for greater accuracy continued for a long time, so that as late as the nineteenth century the author of *Sadratnamala* estimated the circumference of a circle of diameter 10^{18} as follows:

314,159,265,358,979,324 correct to 17 places.

It is clear that in deriving the arctan series, the Kerala School showed both an awareness of the principle of integration and an intuitive perception of small quantities and operations with such quantities. However, they seem to have shied away from developing the methods and algorithms of calculus, being perfectly content with the geometrical approach that their European counterparts would eventually replace with calculus. The reason for this reluctance may be found in the epistemological differences between the two mathematical traditions.

It is instructive to note that Kerala calculus did not have a Greek mathematical epistemology. Possibly influenced by Aryabhata, the Kerala calculus adopted a pragmatic epistemology. The proof paradigm in this epistemology did not place much value on thought experiment, since Indian mathematics tended to be bundled in with the empirical science of astronomy. Indian astronomy had established a "unique epistemology."[19] This "unique epistemology" was shared by Indian mathematics in which proof emphasized "convincing demonstrations,"[20] analogous to the "explanatory notes which served to convince and enlighten" in medieval Chinese mathematics.[21] In Indian mathematics, there was little tension between empirical demonstration (or numerical calculation) and proof. In the *Yuktibhasa*, one sees empirical demonstration or numerical calculation and deduction used without attributing to it a thought experiment of a superior, infallible status.[22] It might be claimed that the achievements of the Kerala mathematics support this perspective as they obtained results in the calculus through pragmatic methods that predate those purportedly obtained by Platonic proof in the Renaissance by several centuries.[23]

TRANSMISSION: CERTAIN METHODOLOGICAL POINTERS

The basis for establishing any transmission of science is often *direct* evidence of records or translations of the relevant manuscripts. The transmission of Indian mathematics and astronomy since the early centuries CE via Islamic scholars to Europe is supported by direct evidence. The transmission of Indian computational techniques was in place by at least the early seventh century, for by 662 CE it had reached the Euphrates region.[24] Indian astronomy was transmitted westward to Iraq by a translation into Arabic of the *Siddhantas* around 760 CE and then into Spain.[25]

Our impetus to search for other types of evidence to "prove" the conjecture of transmission of Kerala mathematics to Europe was initially inspired by a test proposed by Neugebauer, based on the multiple criteria of chronological priority of discovery, availability of accessible communication routes, and the existence of significant methodological similarities.[26] The first criterion regarding *priority* of Kerala calculus is now beyond any doubt.[27] The second is validated by the existence of a long-standing corridor of communication between the Southern India and the Arabian Gulf (via the port of Basrah) for centuries.[28] The arrival of the Portuguese Vasco da Gama to the Malabar Coast in 1499 heralded a more direct route between Kerala and Europe via Lisbon. Thus, despite its geographical location, which prevented easy communication routes with the rest of India, Kerala was linked by sea with the rest of the world and, in particular, Europe.

As far as methodological similarities are concerned, the key result already given in equation (1) above is essentially the same as found in the *Yuktibhasa* and in Kerala mathematics generally. The result assumes a practical knowledge of integration, for it permits the evaluation of the area under the parabola $y = x^k$, or, equivalently, the calculation $\int x^k dx$ And this is precisely what Fermat, Pascal, Wallis, and other European mathematicians did in the seventeenth century to evaluate the area under the parabola. Indeed, the "distance" between formal differentiation and the practical methods deployed in the Kerala calculus is insignificant.[29] Furthermore, our conjecture on methodological similarities between the two mathematical traditions is given credence by the fact that John Wallis used a numerical induction technique not dissimilar to the one noted in the *Yuktibhasa*.[30]

Methodological dissonance could also be a valuable indicator, albeit indirectly, of possible transmission. It is important to recognize that the unconventional way, at least by modern standards, of the passage to infinity in Kerala mathematics shows the existence of an independent tradition of Indian mathematics.[31] The European Renaissance mathematicians, however, worked in the epistemology of Greek mathematics.[32] In this epistemology, the emphasis was on "proof" rather than calculation, and the study of the infinite or the infinitesimal was problematic if not alien.[33] So it may be suggested that the fact that Renaissance mathematicians dealt with the infinite implied non-Greek influences. Here we may cite the fact that Pascal tried to establish the formula (1) cited above by using the so-called Pascal's triangle for the binomial coefficients. This triangle had already appeared in the third-century work *Chandasutra* of Pingala and later in the tenth-century work of the commentator Halayudha. By the eleventh century it was also known to the Islamic world and the Chinese.[34]

THE MODE OF TRANSMISSION: A CONJECTURE

Circumstantial evidence at hand points to the Jesuits as being the possible conduit of transmission of the Kerala mathematics to Europe. The famous Matteo Ricci was in the first batch of Jesuits trained in the new mathematics curriculum introduced in the Collegio Romano by Clavius.[35] He also went to Lisbon to study cosmography and nautical science. Ricci was then sent to India in 1578 and spent a year in Cochin where Jesuits maintained a large presence until the Protestant Dutch closed down the Cochin Jesuit College around 1670. Subsequently, many other Jesuit scientists trained both by Clavius or Grienberger went to India for substantial periods of time. Most notable of these, in terms of their scientific activity in India, were Johann Schreck and Antonio Rubino.[36] The former had studied with the French mathematician Viete, well-known for his work in algebra and geometry. Other Jesuits just as qualified also went to India.[37] And at some point in their stay in India a number were posted to the Malabar region that included the city of Cochin, the epicenter of developments in Kerala mathematics. Further, to not only aid conversions but also to gather local knowledge, the Jesuits became proficient in languages like Malayalam, Telegu, and Tamil.[38]

Ricci and others sought Indian calendrical knowledge.[39] They could not but have noticed the discrepancies between their calendar and the local calendar. For example, the Jesuits were accustomed to the idea that festivals such as Christmas fell on a fixed day of their calendar, so they could not have failed to notice that the major Indian festivals like Dussehra, Diwali, and Holi did not fall on the same days of the Julian calendar each year. However, the Jesuit missionaries prior to Matteo Ricci probably were not sufficiently versed in astronomy to have understood the difference even between the sidereal year (the basis of the Indian calendar) and the tropical year (the basis of the Julian/Gregorian calendar); and so could not have been expected to understand the intricacies of the Indian calendar. However, they could easily have acquired the knowledge in a written form and sent it back to either Maffei in Portugal or to Clavius in the Collegio Romano for analysis. We should emphasize that the Jesuits had much more than a casual interest in the calendar. For at just about that time, Ricci's teacher, Christopher Clavius was a member of the Calendar Reform Committee that was responsible for the introduction of the Gregorian calendar in 1582, an event that had been preceded by centuries of controversy. The 1545 Council of Trent had already acknowledged the error in the Julian calendar, and had authorized the pope to correct it. So Ricci's interest in the Indian calendar was not a casual one, but was an effort preceded by years of preparation and study, and came at a time when the Jesuit interest in both India and the calendar was at a peak. And it is not beyond plausibility that the Patriarch of Antioch, another member of the Calendar Reform Committee who was knowledgeable about the Indian calendar from his Syrian Christian connections, could have enlightened Clavius about the Indian calendar.[40]

Indeed, it appears that the Jesuits tried to formalize this policy by including local sciences such as astrology (or *jyotisa*) in the curriculum of the Jesuit colleges in the Malabar Coast.[41] Furthermore, the early Jesuits were also active in the transmission of local knowledge back to Europe. Some evidence of this knowledge quest is contained in the collections *Goa 38, 46,* and *58* to be found in the Jesuit historical library in the Archivio Pontificia Università Gregoriana (ARSI), Rome. The ARSI collection contains the work of Fr. Diogo Gonsalves on the judicial system, the sciences, and the mechanical arts of the Malabar region. This work started from the very outset of the Portuguese presence in India.

If the early and late Jesuits were involved in learning the local sciences then, given the scientific credentials of the Jesuits such as Ricci of the early period and Schreck and Rubino of the middle period, it is a plausible conjecture that this work was carried on by others. For example, there is de Menses who, writing from Kollam in 1580, reports that, on the basis of local knowledge, the European maps contain inaccuracies.[42] Antonio Rubino wrote, in 1610, similarly about inaccuracies in European mathematical tables for determining time.[43] Then there is the letter from Schreck, in 1618, of astronomical observations intended for the benefit of Kepler—the latter had requested the eminent Jesuit mathematician Paul Guldin to help him to acquire these observations from India to support his theories.[44]

Whilst this does not establish the fact that these Jesuits knew about the existence or even obtained manuscripts containing the Kerala mathematics, it does indicate that their scientific investigations about the local astronomy and calendrical sciences could have led them to an awareness of this knowledge. There are reports that that the Brahmins were secretive and unwilling to share their knowledge.[45] However this was not an experience shared by others. For example, Fr. Diogo Gonsalves, who was mentioned earlier, wrote a book about the administration of justice, sciences, and mechanical arts of the Malabar from information supplied by local informants. Around the same time a Brahmin spent eight years translating Sanskrit works for Fr. Frois.[46]

How might the Jesuits have obtained key manuscripts of Indian astronomy and mathematics such as the *Tantrasangraha* and the *Yuktibhasa*? Clearly such manuscripts would require the Jesuits being in close contact with scholarly Brahmins or Kshatriyas who, in the case of the latter in Kerala were mainly members of the royal family. And the Jesuits working on the Malabar Coast had close relations with the royal family of Cochin,[47] despite their aggressive evangelical work. Furthermore, around 1670, King Rama Varma granted them special privileges that resulted in their greater influence in the royal court.[48] There is considerable evidence that the royal family of Cochin contained a number of notable students of astronomy and mathematics.[49] This cumulative evidence of the continuing tradition of scientific scholarship among some of the royal personages together with the undoubted Jesuit influence in the royal courts of Kerala indicate how the Jesuits could have acquired the knowledge of Kerala mathematics and astronomy.

TESTING DIRECT TRANSMISSION: THE DATA

Prior to the AHRB-funded research project, we had examined the correspondence of Renaissance mathematicians organized by Marin Mersenne.[50] The minim monk Marin Mersenne, through his correspondence with the leading scientists and mathematicians of the early seventeenth century, was an important conduit for transmission of knowledge. In this correspondence, while there is mention of Brahmins,[51] of the works of Orientalists Gaulmin,[52] and Erpen and his "les livres manuscrits Arabics, Syriaques, Persiens, Turcs, Indiens en langue Malaye,"[53] and those of Golius and Drusius at the University of Leyden,[54] there is no specific mention of any Indian mathematicians.

In addition, we had conducted a survey of catalogs in the Vatican library. These indicated the presence of a large number of palm-leaf manuscripts in Malayalam (the language of Kerala) and Tamil (the language of the neighbouring state, Tamil Nadu). It was decided to focus our search to the following mass of materials to determine whether or not there was *direct* evidence of transmission:

1. Archivum Romanicum Societate Iesu, ARSI, Rome. A study of mainly manuscript letters and reports from Jesuit missionaries to their headquarters in Rome.
2. Gregorian University Archives, Rome. A study of manuscript correspondence of scientist Jesuits (amongst them Rubino, Ricci, Schreck) to Clavius and Grienberger.
3. University of Coimbra Archives, Coimbra. An investigation to identify any Jesuit correspondence and materials and to examine the works

by the Jesuit mathematician Borri. He was the only Jesuit missionary who went to the Malabar and returned. He specialized in astrology.
4. Ajuda library, Lisbon. A study of manuscript correspondence of the earlier Jesuit missionaries to the Malabar up to 1568.

Further, in our earlier studies, we noticed a similarity in the approach to certain results in the prototypical calculus of the *Yuktibhasa* and the approach to calculus adopted by some Renaissance mathematicians. This did not come as a surprise to us, since some results of Indian mathematics of five hundred years earlier—notably those of Bhaskara II—were rediscovered in the Renaissance by Fermat and Wallis.[55] So we decided to examine the works of prominent Jesuit mathematicians in Rome who may have analyzed the transmitted manuscripts of the Kerala mathematics. The most prominent of the Jesuit mathematicians of the period in question were the two mentioned previously, Christopher Clavius[56] and Christopher Grienberger.[57] In addition, we know from our earlier studies[58] that scholarly Jesuit missionaries in the Malabar such as Ricci and Rubino were in correspondence with Clavius and Grienberger.

The materials studied included the following:

1. Works and correspondence of Christopher Grienberger. The correspondence was identified from the edited correspondence of Clavius.[59] The mathematical works we identified were his unstudied manuscripts GES 874 and GES 600 in the Biblioteca Nazionale, Rome.
2. Works and correspondence of Christopher Clavius. The correspondence was identified in the edited correspondence as mentioned above and the work identified was the one work that may have been susceptible to influence from Indian sources: *Theodosii Tripolitae Sphaericirum Libri III*, located in the Universitaria Alessandrina in Rome. This deals with the calculation of the trigonometric ratios that we posited to have some connection with the Kerala mathematics.

RESULTS OF THE INVESTIGATION

The materials identified in the previous section were studied in-depth. The net that we cast over the materials was fine and therefore necessarily captured large amounts of information not directly relevant to the aims of the research. For example, the further Jesuit correspondence examined at the ARSI revealed only material related to the missions [conversions, finance, the establishment and administration of the colleges, etc.]. Again, the material at the Ajuda library did not reveal any evidence of information gathering and reporting of mathematical or astronomical nature from the Jesuit missionaries in the Malabar. Examination of the Jesuit documents and correspondence merely confirmed the findings of our prior study that there was strong motivation on the part of the Jesuit missionaries to acquire the science of the Malabar.[60]

We examined the correspondence between Clavius and Grienberger, on the one hand, and the Renaissance mathematicians in contact with them [notably Adrian van Roomen], on the other, and found that they were working in an epistemology that seemed uninfluenced by Kerala mathematics. However, influences of the earlier Indian mathematics notably from that of Aryabhata and his early commentators were detected. In this context it should be pointed out that Clavius was essentially rewriting the earlier work on trigonometric tables by Regiomontanus and thus it will be truer to say that the latter rather than Clavius was influenced by Aryabhata. For in Regiomontanus's *Compostio Tabularum suinuum Rectorum*, published in 1541 he, in the manner of Ptolemy, calculates the sines of 90°, 45°, 60°, 30°, and 15° from appropriate triangles, and then uses the right-angle theorem to calculate sine 75°. He then uses the formula, "For a random arc in a quadrant, the sine is the middle proportional [geometric mean] between half the radius and the versed sine of double the arc" that is equivalent to Aryabhata's *kramatkramajya* rule.[61] This finding is in accordance with Knobloch's work[62] on the Islamic influence on Clavius's work, as we note that the Arab mathematics of the Islamic world was itself influenced by developments in India[63].

Grienberger's hitherto unstudied manuscript works GES 674 and GES 600 were examined in detail. It revealed a lengthy work on the computation of trigonometric tables graduated in degrees with an intended accuracy of 18 places of decimals. The novel methods and algorithms used in the lengthy calculations were deciphered with some difficulty. It revealed that Grienberger's methods were novel, utilizing methods from many sources, but there was no direct evidence of influence, although the arithmetic used was Indo-Arabic and the methods of construction of the tables showed an Aryabhatan flavor. We had initially conjectured that the initial value of sine of 1 minute [correct to 22 places of decimals] used by Grienberger to generate his tables was calculated using the infinite series for sine that had been discovered earlier by the Kerala mathematicians. However, our analyses made it clear that the infinite series of trigonometric ratios is impracticable for the construction of trigonometric tables graduated in degrees.

TENTATIVE CONCLUSIONS

The painstaking trawl of the mass of manuscript and other materials has yielded no *direct* evidence of the conjectured transmission. We have to therefore report that on the basis of the evidence of the documents studied so far, the European Renaissance developments of early calculus may well have been independent of the developments in that subject in Kerala some centuries earlier. This can only be a provisional conclusion as it is by no means the case that all of the material that is required to be studied to reach a definitive conclusion has been studied. There may be materials available amongst the mass of uncataloged documents in Portugal or in private libraries in Italy. Both these countries suffered upheavals in which library contents were dispersed—in Portugal the suppression of the Jesuits in 1750 by the Marquis of Pombal and in Italy the effects of the Thirty Years' War in the early seventeenth century.

In conclusion, it is worth noting that we have focused so far only on evidence of *direct* transmissions (through written records) of Kerala ideas to Europe. But, as pointed out by Bala the transmission of the discoveries of Kerala mathematics could have been as "know-how" and computation techniques through the channel of craftsmen and technicians.[64] This may explain the absence of documentary evidence in Jesuit communications. However, as Arthur argues it may well be that one can "establish a precedence as well as a strong probability that an influence occurred without being able to find concrete evidence for it."[65] Confining oneself to the view that transmissions occur through "one person reading others work and becoming influenced as a result" may be taking an excessively narrow view of how knowledge travels across cultures. The issue of how mathematical ideas travel remains an intriguing and open question.

END NOTES

* This chapter is based on recent work by Dennis Almeida and the author (Almeida and Joseph, 2004, 2007, 2009) that were the outcomes of their collaboration on a research project funded by Arts and Humanities Research Board (AHRB), United Kingdom. The author owes a considerable debt to Dennis Almeida for his contributions that were crucial to the success of the project. I have reprised short passages from the publications referred to above because, needing to make the same points, I found the way they were expressed there as apt for the task as I could make them.

1. See Boyer (1949), Baron (1969), and Edwards (1979).
2. Although Baron (1969), Calinger (1999), Jushkevich (1964), and Katz (1992) recognize the Indian contributions, Fiegenbaum (1986) mentions a group of European mathematicians who anticipated the famous "Taylor's theorem" but makes no reference to the Kerala work. And this is despite the easily accessible discussion on Kerala mathematics in Rajagopal and his various collaborators (1944, 1949, 1952, 1978, and 1986), as well as histories of Indian mathematics detailing the Kerala contribution such as Bag (1979), Sarasvati Amma (1979), and Srinivasiengar (1967).
3. See Whish (1835), Rajagopal and Iyer (1952), Sarasvati Amma (1979), Jushkevich (1964), Srinivasiengar (1967), Rajagopal and Rangachari (1978), and Sarma (1972).

4. See *Tantrasangraha* V.53–54 and *Sphutanirnaya* III.19–20. For a recent survey of these and other discoveries by the Kerala School, see Plofker (2009, 217–251) and the two volume-translation of *Yuktibhasa* by Sarma (2008).

5. There is fragmentary evidence to indicate that some members of the Kerala School were younger sons. This fact could be significant. The Nambuthris followed a strict system of primogeniture, so that all landed property went to the eldest son. There was an additional twist. Only the eldest son could marry a Nambuthri woman. The younger sons formed sexual partnerships with Nair women, the children of whom, within the Nair matrilineal system, were the sole responsibility of the Nair mothers. So we had a situation where a number of Nambuthris, freed of all economic and family responsibilities, became a truly leisured class. For further details of the socioeconomic climate of that time, see Mallayya and Joseph (2009a, 35–58) and Vijaylakshmy and Joseph (2009, 59–73).

6. There has been a long Indian tradition of providing *yuktis* (rationales) and *upapattis* (proofs) in commentaries rather than in the main text that often contained results expressed in cryptic verses that could be easily memorized.

7. The role of the temple as an institution creating and sustaining intellectual activities needs further study. As a meeting place of those involved in the study of mathematics and astronomy, as a vehicle for receiving and disseminating scientific knowledge, as an agency for recruiting able students and practitioners outside the confines of narrow caste and regional lines, in all these cases the temple may well have played an important role. For further discussion of the role of the temple in the economic and social life of medieval Kerala, see Gurukkal (1992).

8. This is a slight modification of the translation by K. V. Sarma of the *Yuktibhasa* (2008, 68–69).

9. The first English translation of this seminal text came out in 2008. It was by the doyen among scholars of the history of Kerala mathematics, Professor K. V. Sarma. Unfortunately, Professor Sarma died before its publication.

10. The *Katapayadi* was a system devised to help memorization, since memorable words can be made up using different chronograms. For example, if such a system is applied to English, the letter b, c, d, f, g, h, j, k, l, m would represent the numbers zero to nine. So would n, p, q, r, s, t, v, w, x, y. The last letter, z, denotes zero. The vowels, a, e, i, o, u are helpful in forming meaningful words but have no numerical values associated with them. Thus, according to this notational scheme, the word "Madras" represents 9234 and "love" indicates 86.

11. For a recent exposition and derivation, see Mallayya and Joseph (2009b, 89–95).

12. "Approximately" is an insufficiently precise translation of the Sanskrit work *asana*, since it does not contain the notion of "unattainability."

13. Section 3 of Chapter 6 of the *Yuktibhasa* is entitled "Circumference of a Circle without Calculating Square Roots." And the subsection is titled "Dividing the Circumference into Arc-Bits: Approximating the Arc-Bits by *Jyardha-S* (Rsines)" (Sarma 2008, 49).

14. However, the most likely direct source is verse 26 of the *Aryabhatiya*.

15. A familiar result from elementary algebra is that the series $1-x+x^2-x^3+\ldots$ represents for $x<1$ an infinite decreasing geometric progression with common ratio $-x$ and sum equal to:
$$1 = 1 = (1+x)-1$$
$$1-(-x)1+x$$

16. This asymptotic relation was first noted in Europe by Gilles de Roberval in 1634 and Pierre de Fermat in 1636.

17. For a useful discussion of these approaches, see Hayashi et al. (1990).

18. The correction used involved incorporating the following as the last term: $F_c(n) = (n^2 + 1)/(4n^3 + 5n)$, where n is the number of terms

19. Pingree (1981).

20. Joseph (1994).

21. Man-Keung Siu (1993).

22. From a pedagogical point of view, it may be suggested that any tension that exists between the means by which one obtains reliable and valid knowledge in Indian mathematics and the proof tradition used in formal mathematics today is analogous to the tension between the informal practices of mathematics learners and the formal practices their teachers would like them to inculcate.

23. Baron (1969, 62) wrote:

    The Hindus seem to have been attracted by the computational aspect of mathematics, partly for its own sake and partly as a tool in astrological prediction. No trace has been found of any proof structure such as that established by Euclid for Greek mathematics nor is there any evidence which suggests Greek influence. Nevertheless, some of the results achieved in connection with numerical integration by means of infinite series anticipate developments in Western Europe by several centuries.

24. See Berggren (1986, 30).

25. See Subbarayappa and Sarma (1985, 38).

26. Neugebauer (1962, 66–67).

27. This has been noted by Baron (1969, 63) and Katz (1995, 173–174).

28. Potache, 1989.

29. Rajagopal and Marar (1944, 65–72) wrote:

    There are two points which emerge from a consideration of the [mathematics of the Kerala School]...In the first place, it employs relations which would appear not to have been noticed in Europe before modern forerunners and followers of the calculus started investigations . . . Our second point is not unconnected with the first. The Hindu mathematicians achieved, without the aid of calculus, results which, for us, are treated best by means of the calculus...This is not to gainsay the fact that (i) the Hindus' proof of [infinite] series shows their awareness of the principle of integration as we ordinarily use it nowadays (ii) their intuitive perception of small quantities like $O(1/n^p)$, $n \to \infty \ldots$ is as good as a practical knowledge of differentiation.

30. For Wallis' "induction" method, see Scott (1981, 30) and Jesseph (1999, 42).

31. See Colebrooke (1873, 446), vol. 2; Pingree, (1981, 8–9).

32. Dantzig (1955).
33. See Boyer (1949, 121–122 and 221).
34. Katz, 1995, 164.
35. Cronin (1984, 22) states that "in 1575 Ricci entered a new phase of his studies; philosophy and mathematics, Aristotle and Euclid. The advanced course was taught by a young German, Christopher Clavius, the most brilliant mathematician of his day...He showed special aptitude for this course winning notice as a mathematician of promise."
36. See Baldini (1992) and Iannaccone (1998).
37. Baldini (1992, 70) writes:

   Si può ricordare che molti dei migliori allievi gesuiti di Clavio e Geienberger (iniziando con Matteo Ricci e proseguendo con C. Spinola, G. Aleni, G. A. Rubino, S. De Ursis, Schreck, G. Rho) divennero missionari nelle Indie orientali. Questa scelta, se li fece protagonisti di un interscambio tra la tradizione europea e quelle indiana e cinese, particolarmente in matematica ed astronomia, che fu di per sé un fenomeno di grande significato storico, limitò certamente la loro produttività scientifica.

   Translation: It can be recalled that many of the best Jesuit students of Clavius and Grienberger (beginning with Ricci and continuing with Spinola, Aleni, Rubino, Ursis, Schreck, and Rho) became missionaries in Oriental Indies. This made them protagonists of an interchange between the European tradition and those Indian and Chinese, particularly in mathematics and astronomy, which was a phenomenon of great historical meaning.

38. See, for example, Correia-Afonso (1997, 47, 64), Correia-Afonso (1969, 58, 92), and Wicki (1948–, XV, 34).
39. In a letter to Maffei, he states that he requires the assistance of an "intelligent Brahmin or an honest Moor" to help him understand the local ways of recording and measuring time or *jyotisa*. See Venturi (1913, 24).
40. Grafton (1993, 108).
41. See, for example, Wicki (1948–, vol. 3, 307).
42. "I have sent Valignano a description of the whole world by many selected astrologers and pilots, and others in India, which had no errors in the latitudes. For the benefit of the astrologers and pilots that every day come to these lands, their maps are all wrong in the indicated latitudes..." (Wicki 1948–, vol. XI, 185).
43. "Comparing the real local times with those inferable from the ephemeridis [tables] of Magini, he [Rubino] found great inaccuracies and, therefore, requested other ephemeridi" (Baldini 1992, 214).
44. Baumgardt (1951, 153), Iannaccone (1998, 58).
45. This is noted by D'Elia (1960, 15).
46. Ferroli (1939, 402).
47. See for example, Wicki (1948–, vol. X, 239, 838) (vol. XV 591).
48. Wicki (1948, vol. XV, 224).
49. For further details and evidence of the scholarship in the Cochin court, see Almeida and Joseph (2009, 256–275).

50. *Correspondance du P. Marin Mersenne.* 1945–. 18 vols. Paris: Presses Universitaires de France.
51. *Correspondance du P. Marin Mersenne,* vol 13, 518–521; a letter from Mersenne to Buxtorf. Mersenne mentions the knowledge of Brahmins and "Indicos."
52. *Correspondance du P. Marin Mersenne,* vol. 13, 518–521.
53. Ibid., vol. 2, 103–115.
54. Ibid., vol. 2, 155; a letter from Mersenne to Rivet.
55. For details see, The Aryabhata Group (2002, 33–48).
56. The position of Clavius in the history of science is assured by his role in the Gregorian calendar reform and his many publications. Clavius also maintained contacts with scientists and mathematicians outside Rome. Furthermore, Clavius was influenced by Arab mathematicians. For further details, see Knobloch (2001).
57. According to Gorman (2003), Grienberger was "a reviser of mathematical works written by Jesuits and in his strategies of engagement in epistolary relationships with natural philosophers and mathematicians outside the Jesuit order."
58. See The Aryabhata Group, (2002, 33–48).
59. See Baldini and Napoletani (1992).
60. For details see Almeida and Joseph (2004, 45–60)
61. See Shukla and Sarma (1976)
62. See Knobloch (2001)
63. See, for example, Rashed (1994).
64. Bala (2009, 155–179).
65. See Arthur's article in this volume.

BIBLIOGRAPHY

Almeida, D., and G. G. Joseph. 2004. "Eurocentrism in the History of Mathematics: The Case of the Kerala School." *Race and Class* 45, no. 4: 45–60.

———. 2007. "Kerala Mathematics and Its Possible Transmission to Europe." *Philosophy of Mathematics Education Journal,* retrieved from www.people .ex.ac.uk/PErnest/pome20/index.htm.

———. 2009. "A Summary Report of the Investigation on the Possibility of the Transmission of the Medieval Kerala Mathematics to Europe." In *Kerala Mathematics: Its History and Possible Transmission,* edited by G. G. Joseph, pp. 256–275. Delhi: B. R. Publishing Corporation.

The Aryabhata Group. 2002. "Transmission of the Calculus from Kerala to Europe." In *Proceedings of the International Seminar and Colloquium on 1500 Years Of Aryabhateeyam,* pp. 33–48. Kochi: Kerala Sastra Sahitya Parishad.

Bag, K. 1979. *Mathematics in Ancient and Medieval India.* Varanasi: Chaukhamba Orientalia.

Bala, A. 2009. "Establishing Transmissions: Some Methodological Issues." In *Kerala Mathematics: Its History and Possible Transmission to Europe,* edited by G. G. Joseph, pp. 155–180. Delhi: B. R. Publishing Corporation.

Baldini, U. 1992. *Studi su filosofia e scienza dei gesuiti in Italia 1540–1632.* Firenze: Bulzoni Editore.

———, and P. D. Napoletani. 1992. *Corrispondenza/Christoph Clavius; edizione critica a cura.* Pisa: Università di Pisa, Dipartimento di matematica.

Baron, M. E. 1969. *The Origins of the Infinitesimal Calculus.* Oxford: Pergamom.

Baumgardt, C. 1951. *Johannes Kepler: Life and Letters.* New York: Philosophical Library.

Berggren, J. 1986. *Episodes in the Mathematics of Medieval Islam.* New York: Springer Verlag.

Boyer, C. B. 1949. *The History of the Calculus and Its Conceptual Development.* New York: Hafner Publishing Company.

Calinger, R. 1999. *A Contextual History of Mathematics to Euler.* Upper Saddle River, NJ: Prentice Hall.

Colebrooke, H. 1873. *Miscellaneous Essays,* vol. 2. London: W. H. Allen.

Correia-Afonso, J. 1997. *The Jesuits in India, 1542–1773: A Short History.* Gujarat: Gujarat Sahitya Prakash.

———. 1969. *Jesuit Letters and Indian History: 1542–1773.* Bombay: Oxford University Press.

Cronin, V. 1984. *The Wise Man from the West – Matteo Ricci and His Mission to China.* London: Collins.

Dantzig, T. 1955. *The Bequest of the Greeks.* London: Allen and Unwin.

D'Elia, P. 1960. *Galileo in China Relations through the Roman College between Galileo and the Jesuit Scientist-Missionaries (1610–1640),* translated by R. Suter and M. Sciascia. Cambridge, MA: Harvard University Press.

Edwards, H. 1979. *The Historical Development of the Calculus.* New York: Springer-Verlag.

Ferroli, D. 1939. *The Jesuits in Malabar,* 2 vols. Bangalore: Bangalore Press.

Fiegenbaum, L. 1986. "Brook Taylor and the Method of Increments." *Archive for History of Exact Sciences* 34, no. 1: 1–140.

Gorman, M. 2003. "Mathematics and Modesty in the Society of Jesus the Problems of Christoph Grienberger (1564–1636)." *The New Science and Jesuit Science: Seventeenth Century Perspectives,* Archimedes vol. 6, edited by Mordechai Feingold, pp. 1–120. Dordrecht: Kluwer.

Grafton, A. 1993. *Joseph Scaliger: a Study in the History of Classical Scholarship.* Oxford: Clarendon Press.

Gurukkal, R. 1992. *The Kerala Temple and Early Medieval Agrarian System.* Sukupuram: Vallathol Vidyapeetham.

Hayashi, T., T. Kusuba, and M. Yano. 1990. "The Correction of the Madhava Series for the Circumference of a Circle." *Centaurus* 33: 149–174.

Iannaccone, I. 1998. *Johann Schreck Terrentius.* Napoli: Instituto Universitario Orientale.

Jesseph, D. M. 1999. *Squaring the Circle: The War between Hobbes and Wallis.* Chicago, IL: University of Chicago Press.

Joseph, G. G. 2000. *The Crest of the Peacock: Non-European Roots of Mathematics.* Princeton, NJ, and Oxford: Princeton University Press.

———. 1994. "Different Ways of Knowing: Contrasting Styles of Argument in Indian and Greek Mathematical Traditions." In *Mathematics, Education and Philosophy,* edited by P. Ernest, pp. 194–204. London: Falmer Press.

Jushkevich, A. P. 1964. *Geschichte der Matematik im Mittelater.* Leizig: German Translation.

Katz, V. J. 1995. "Ideas of Calculus in Islam and India." *Math Magazine* 68, no. 3: 163–174.

————. 1992. *A History of Mathematics: An Introduction.* New York: Harper Collins.

Knobloch, E. 2001. *Christoph Clavius (1538–1612) and His Knowledge of Arabic Sources,* retrieved from http://www.ethnomath.org/resources/knobloch .pdf.

Mallayya, V. M., and G. G. Joseph. 2009a. "Indian Mathematical Tradition: The Kerala Dimension." In *Kerala Mathematics: History and Its Possible Transmission to Europe,* edited by G. G. Joseph, pp. 35–58. Delhi: B. R. Publishing Corporation.

————. 2009b. "Kerala Mathematics: Motivation, Rationale and Method." In *Kerala Mathematics: History and Its Possible Transmission to Europe,* edited by G. G. Joseph, pp. 89–95. Delhi: B. R. Publishing Corporation.

Mersenne, Marin. 1945. *Correspondance du P. Marin Mersenne,* 18 vols. Paris: Presses Universitaires de France.

Neugebauer, O. 1962. *The Exact Sciences in Antiquity.* New York: Harper.

Pingree, D. 1981. *Jyotisástra: Astral and Mathematical Literature (A History of Indian Literature 6.4).* Wiesbaden: Harrassowitz.

Plofker, K. 2009. *Mathematics in India.* Princeton, NJ, and Oxford: Princeton University Press.

Potache, D. 1989. "The Commercial Relations between Basrah and Goa in the Sixteenth Century." *STUDIA,* Lisbon, 48.

Rajagopal, C. T., and T. V. Iyer. 1952. "On the Hindu Proof of Gregory's Series." *Scripta Mathematica* 18: 65–74.

Rajagopal. C. T., and K. M. Marar. 1944. "On the Hindu Quadrature of the Circle." *Journal of the Royal Asiatic Society (Bombay Branch),* 20: 65–72.

Rajagopal. C. T., and M. S. Rangachari. 1978. "On an Untapped Source of Medieval Keralese Mathematics." *Archive for History of Exact Sciences* 18: 89–102.

Rajagopal, C. T., and M. S. Rangachari. 1986. "On Medieval Keralese Mathematics." *Archive for History of Exact Sciences* 35: 91–99.

Rajagopal, C. T., and A. Venkataraman. 1949. "The Sine and Cosine Power Series in Hindu Mathematics." *Journal of the Royal Asiatic Society of Bengal* 15: 1–13.

Rashed, R. 1994. "Indian Mathematics in Arabic." In *The Intersection of History and Mathematics,* edited by C. Sasaki, J. W. Dauben, and M. Sugiura, pp. 143–148. Basel, Boston, and Berlin: Birkhaüser Verlag.

Sarasvati Amma, T. A. 1979. *Geometry in Ancient and Medieval India.* Delhi: Motilal Banarsidass.

Sarma, K. V., trans. 2008. *Ganita-Yukti-Bhasa (Rationales in Mathematical Astronomy) of Jyesthadeva,* 2 vols. Delhi: Hindustan Book Agency.

————. 1972. *A History of the Kerala School of Hindu Astronomy.* Hoshiarpur: Hoshiarpur Vishveshvaranand Institute.

Scott, J. F. 1981. *The Mathematical Work of John Wallis.* New York: Chelsea.

Shukla, K., and K. V. Sarma. 1976. *Aryabhatiya of Aryabhata.* New Delhi: Indian National Science Academy.

Siu, Man-Keung. 1993. "Proof and Pedagogy in Ancient China: Examples from Liu Hui's Commentary on Jiu Zhang Suan Shu." *Educational Studies in Mathematics* 24: 345–357.

Srinivasiengar, C. N. 1967. *The History of Ancient Indian Mathematics.* Calcutta: Calcutta World Press.

Subbarayappa, B. V., and K. V. Sarma. 1985. *Indian Astronomy – A Source-Book.* Bombay: Nehru Centre Publications.

Venturi, Tacchi. 1913. *Matteo Ricci S.I., Le Lettre Dalla Cina 1580–1610*, vol. 2. Macerata: Stab. Tip. Giorgetti.

Vijaylakshmy, M., and G. G. Joseph. 2009. "An Intellectual History of Medieval Kerala with special Reference to Mathematics and Astronomy." In *Kerala Mathematics: History and Its Possible Transmission to Europe*, edited by G. G. Joseph, pp. 59–73. Delhi: B. R. Publishing Corporation.

Whish, C. 1835. "On the Hindu Quadrature of the Circle and the Infinite Series of the Proportion of the Circumference to the Diameter Exhibited in the Four *Shastras*, the *Tantrasamgraham*, *Yukti-Bhasa*, *Carana Padhati*, and *Sadratnamala*." *Transactions of the Royal Asiatic Society of Gr. Britain and Ireland* 3: 509–523.

Wicki, J. 1948–. *Documenta Indica*, 16 vols. Rome: Monumenta Historica Soc. Iesu.

3

COPERNICUS, ARABIC SCIENCE, AND THE SCIENTIFIC (R)EVOLUTION

Michal Kokowski

INTERCULTURAL DIALOGUE AND SCIENTIFIC (R)EVOLUTIONS

According to some scholars of Arabic science, and their adherents, Nicolas Copernicus of Warmia—a part of Royal Prussia under the Crown of the Polish Kingdom—was not an original, revolutionary scholar. They claim that he simply applied the best methods and results achieved by ancient Hellenic and Hellenistic science, and Medieval Arabic science (from both the Eastern and the Western traditions) in his works, but did not add, in principle, any new ideas. There are several variants of this thesis. For example, Copernicus has been treated as a "European satellite of the Maragha astronomers" by Peter Barker and Roger Ariew, as "the last heir of Arabic medieval astronomy" by Henri Hugonnard-Roche, and as simply continuing "a Maragha revolution in science" by George Saliba. Such views lead them to conclude that "there was no Copernican revolution in science."[1]

However, these views can be challenged because the opponents of Copernicus's originality assume oversimplified ideas of what constitute a scientific method and a scientific revolution. In this chapter, I will defend Copernicus's originality by drawing upon the notion of a *"scientific (r)evolution."*[2]

To appreciate this notion, let us look at the view shared by many educated people that the progress of science occurs in a two-stage process. First, science develops in an evolutionary manner and then a scientific revolution takes place. This scientific revolution—in an analogy to political revolutions, such as the Russian revolution or the French revolution—is seen as completely displacing the earlier tradition in science. Through the course of such revolutions, new fundamental theories, and even scientific method, are perceived to be so radically changed that they become incommensurable with their predecessors. It leads to the perception that the greatest scholars, these being the authors of revolutionary discoveries, should be perceived as adversaries of a creative dialogue between the old and the new science.

Such views stem from the works of Thomas S. Kuhn (1922–1996) and Paul K. Feyerabend (1924–1994), historians and philosophers of science.[3] These authors advocate a radical discontinuity thesis for the progress of science. However, such a thesis is definitively wrong because the concepts of a "scientific evolution" and a "scientific revolution" are only idealizations of the actual development of science.[4] Indeed even the best scientific theories, such as the theory of gravitation and Newtonian mechanics, the theory of special relativity, the theory of general relativity and quantum mechanics, that define famous scientific revolutions (such as the Newtonian revolution, the relativistic revolution, and the quantum revolution) accepted many ideas and views from previous theories including their empirical bases to a large extent. Thus, they also had an evolutionary nature.

One way of seeing this process is to view these new theories as linked with the old ones by means of so-called generalized correspondence principles, that is, scientists try to save all the possible old views that are well-founded on both empirical data and on theoretical ideas by engaging in *a creative dialogue* between the established science (established scientific truths) and the new science (that is not yet established well). In effect, scientists look for new, more general theories than the previous ones. We say in this context that scientists are directed in this search by the postulate of correspondence of theories. In other words, scientists look for a new theory that does not lead to *an arrogant monologue* that rejects all established science.

To illustrate the view presented of the general correspondence principle, it is worth quoting the graphs of the dependence of normalized inertial mass (that is the quotient of inertial mass and inertial mass for zero speed) according to relativistic mechanics (RNM) and classical mechanics (CNM) (See figure 3.1).[5]

It is necessary to add one comment here. The theories that were mentioned assumed different quasi-ontological, logical, and mathematical structures. In this sense, the new theory (relativistic mechanics) rejected quasi-ontological, logical, and mathematical structures of the old theory (classical mechanics). Nevertheless, relativistic mechanics had many elements (both theoretical ideas and empirical data) in common with classical mechanics. For example, relativistic mechanics accepts the old data assumed by classical mechanics, in the limited range of speed that is low in comparison to the speed of light. Hence, even the most revolutionary theory has many components of an evolutionary origin.

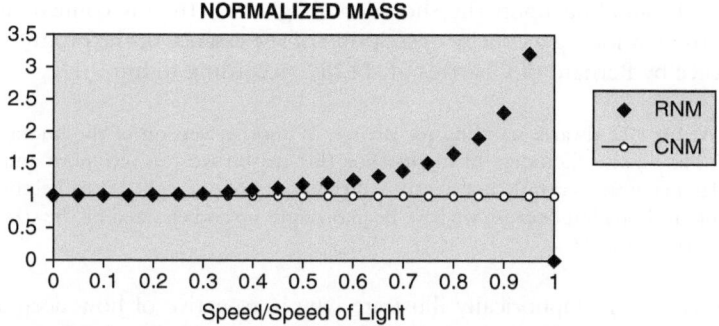

Figure 3.1 Graph of the dependence of normalized inertial mass.

For reasons of this kind, I state that the greatest events in the development of science, the cornerstones of the progress of science, that are described in literature as *scientific revolutions*, are in fact *scientific (r)evolutions*—as sketched above.[6]

Furthermore, I think that one of the best indicators of the maturity of a branch of science and of a scientific (r)evolution having occurred, is the following criterion. The formulation of a new (scientific) theory/law is linked with its predecessor theory/law by means of a generalized correspondence principle. The notion of a generalized correspondence principle can be understood as follows. In their scientific dialogue with nature in mathematical language, scientists look for a new, more general theory to replace earlier ones. But this dialogue always takes into account the content of the old science, the previous theories. It is more like a translation and/or transformation and/or reproduction of the previous knowledge, both the old empirical data and the old theoretical notions (even including all the constructs of these theories) in a wider context.[7]

The classical examples of theories linked by a generalized correspondence principle are quantum mechanics and classical mechanics, or relativistic mechanics and classical mechanics that define two famous scientific (r)evolutions: quantum and relativistic. This strategy of "a dialogue with nature in mathematical language"[8] can be called a "hypothetico-deductive method of correspondence-oriented thinking." It combines the hypothetico-deductive method with the method of correspondence-oriented thinking. In adopting this method, one formulates a new mathematical hypothesis that in certain limiting contexts approximates mathematical regularities predicted by the predecessor hypothesis—that is, the new mathematical predictions "correspond" to the predictions of the earlier theories in limiting cases.[9]

It allows us to see the progress of science through scientific (r)evolutions, which constitute milestones in the progress of science, as one of "dwarfs standing upon the shoulders of giants." In this context, it is worth mentioning a concise description of the essence of the progress of science by Bernard of Chartres (d. 1128). According to him:

> We are like dwarfs standing (or sitting, in another version of the quotation) on the shoulders of giants. For this reason we can see more and farther away, certainly not because of the acuity of our sight or the height of our body but because we have been brought up and elevated by the size of the giants.[10]

It serves to metaphorically illustrate why irrespective of how deep scientific (r)evolutions occur their authors are not radical scientific revolutionaries in T. S. Kuhn's sense. They are only scientific (r)evolutionaries, since they always proclaim only partly new radical ideas. This is best depicted by the Greek myth of Cedalion standing on the shoulders of Orion. In the myth Cedalion, or Kedalion, was a servant of Hephaestus of Lemnos. The myth tells of how Cedalion brought about the healing of Orion's sight after he was blinded by Oenopion and came to Lemnos. Orion set Cedalion upon his shoulders to guide him on a journey toward the East, where the rays of Helios restored his sight.

Why do I mention this myth in the context of discussing the "progress of science"? I think it supplements very well Bernard of Chartres's metaphoric picture of "dwarfs standing upon the shoulders of giants." Each scientist—even the greatest—is always a dwarf, Cedalion, on the shoulders of a blind giant, Orion, who represents the tradition determined by the achievements of all previous scientists. I say "blind," since the old science cannot show the new directions, being closed to the crucial ideas and discoveries of the new science. To find them, as Cedalion, we must have very good sight, in the literal and/or metaphoric sense, and as Cedalion be able to guide the old science ("blind Orion") to new knowledge ("the rising Sun that restores Orion's sight").

COPERNICUS'S THOUGHT AND ARABIC SCIENCE IN A DIALOGICAL PERSPECTIVE

Copernicus's astronomical models, Ptolemy's astronomy and medieval Arabic astronomy

Since prehistoric times, astronomy has been strictly linked with mythical and religious thought. The periodicity of the most important astronomical phenomena (such as vernal and autumnal equinoxes, summer and winter solstices) have often served to determine the most important dates for religious rituals.[11] One can find the expression of such links between astronomical science and religion in many professional notebooks of Ancient, Medieval, and Renaissance astronomy, written in Africa, Asia, and Europe. These include, the *Almagest* by Claudius Ptolemy (90–168) from Alexandria (Roman Egypt, Africa); *al-Tadhkira fi `ilm al-hay' a (A memoir on the science of astronomy)* by Nasīr al-dīn al-Tūsī (1201–1274), a Persian of the Ismaili, who lived, among other places, in Baghdad (Iraq, Asia);[12] and *De revolutionibus (On the revolutions)* by Nicholas Copernicus (1473–1543).[13]

Indeed, irrespective of whether we are in eclectic Roman Egypt in the second century, the Islamic Empire (in Asia, Africa, or Europe) in the thirteenth century, or in Catholic Warmia, under the Crown of the Polish Kingdom (a Kingdom tolerant of different religions), in the sixteenth century, we read similar words praising God as the source of Truth and as the Creator of the ordered universe. This link with astronomical science is common to many religions as shown in figure 3.2.

The close link between astronomical phenomena and the fixing of auspicious religious events also led to careful observation of the heavens. But such data, even into early modern times, were based on ancient observations. Copernicus accepted Ptolemy's observations reported in the *Almagest* that also include data taken from Hipparchus (about 150 BC) that virtually originated in ancient "Babylonian" astronomy of

Figure 3.2 Copernicus's conversations with God (1873) by Jan Matejko (1838–1893), Cracow, Jagiellonian Library.

the Hellenistic period from the fourth to the second centuries BC.[14] But although later professional astronomers of medieval times and the Renaissance made astronomical observations with the same precision as Ptolemy, those observations differed from the observations of Ptolemy and his predecessors. This was obvious to Copernicus from his reading of professional medieval and Renaissance astronomical works, including the Parisian version of the *Alphonsine Tables* (written in the 1320s, and first printed in Venice in 1483), and *Epitome in Almagestum Ptolemei* written by Georg Puerbach (1423–1461) and Joannes Regiomontanus (1436–1476) (printed in Venetia in 1496). This leads Copernicus to conclude in his *Commentariolus, Letter against Werner* and *De revolutionibus*, that Ptolemy's astronomical theory could not save astronomical phenomena over a long span of time (that is of several hundred years).[15]

However, all such works were based on the achievements of Arabic astronomy, which flourished from the eighth to the fifteenth centuries throughout Arabic culture from the East, in Asia, to the West, in Spanish Europe.[16] Hence, it is no accident that Copernicus accepted observations and parameters of astronomical models from a number of Arabic scholars. These include Al-Sābi' Thābit ibn Qurra al-Harrānī (born in 836 in Harran, Mesopotamia—died in 901 in Baghdad); Muhammad ibn Jābir al-Harrānī al-Battānī (born c. 858 in Harran—died in 929 in Qasr al-Jiss, near Samarra; he worked in Syria, at ar-Raqqah and at Damascus; Abū Ishāq Ibrāhīm ibn Yahyā al-Naqqāsh al-Zarqālī, Latinized as Arzachel (1029–1087); he flourished in Toledo in Castile, Andalusia [now Spain]); Isaac ben Sid and Judah ben Moses ha-Cohen, the Jewish authors of the *Toledan Tables* (c. 1070), translated into Latin by Gerard of Cremona in the twelfth century; Nur ad-Din al-Betrugi (Latinized as Alpetragius) (he was born in Morocco, settled in Seville, in Andalusia, and died c. 1204); and the authors of the Parisian version of the *Alphonsine Tables*.

Furthermore, in Copernicus's works we can also find a set of parameters of several astronomical models (and even geometrical constructions) that appear to descend from the scholars of the Maragha school (Maragha, Northern Iran), including Nasīr al-Dīn al-Tūsī (born 1201 in Tūs, Khorasan—died 1274 in al-Kāzimiyyah, Baghdad), and Ala Al-Din Abu'l-Hasan Ali Ibn Ibrahim Ibn al-Shatir (1304–1375) of Damascus (Syria). However, it is still a matter of controversy how Copernicus could have acquired knowledge of these achievements. It occurred, possibly, through his careful reading of several works printed in Padua or in Vienna.[17]

It is worth noticing that Copernicus constructed two versions of the astronomical system with a mobile Earth. The first version, described in the *Commentariolus* (written about 1508), is a system that saves short-period astronomical phenomena using observational data taken from the Parisian version of the *Alphonsine Tables* (written in the 1320s; first printed in Venetia in 1483), by assuming constant inclination of the equator, and constant speed of precession of the equinoxes. The final version, described in the *De revolutionibus* (printed in 1543), saves both short- and long-period astronomical phenomena with periodically changed speed of the precession of equinoxes, and periodically changed inclination of the equator,[18] based on observational data taken from Ptolemy's *Almagest* and later works such as the Parisian version of the *Alphonsine Tables*, Georg Puerbach's and Joannes Regiomontanus's *Epitome in Almagestum Ptolemei*, and on his own observations made from 1497 to 1541.

In the process of building these systems, Copernicus not only accepted a great deal of observational data from Arabic astronomy, but also many theoretical elements, including the so-called Tusi's device or Tusi couple (See figure 3.3).[19] He applied the device mentioned, among others, to remove Ptolemy's geometrical construction of an equant as it contradicted the so-called Plato's principle (according to which all astronomical motions must be explained by a set of uniform circular motions around physically real, geometrical centers of circles).

The latter point is of great importance.[20] Doing so, he agreed with the critical estimation of Ptolemy's astronomy by Arabic astronomers who had also invoked Plato's principle to criticize Ptolemy's use of the notion of equants. [21] These Arabic critics include, among others, Ibn al-Haytham (known in the West by the Latinized name of Alhazen; born in 965 in Basra, Iraq, and part of Buyid Persia at that time; died c. 1039 in Cairo)[22]; Ibn Rushd, better known as Averroes in European literature (born in Córdoba, modern-day Spain, in 1126; died in 1198 in Marrakech, modern-day Morocco); and Nur ad-Din al-Betrugi (known in the West by the Latinized name of Alpetragius, born in Morocco, he settled in Seville, in Andalusia; died ca. 1204), and the astronomers linked with the Maragha school.

Nevertheless, Copernicus's decisions to accept many theoretical elements from Arabic astronomy, including the Tusi couple, were taken by many scholars to imply that Copernicus was rather an unoriginal scholar.[23] Some more radical scholars assume that even all the important theoretical elements that determined Copernicus's achievements, including the idea of heliocentrism and a heliocentric system, were borrowed from Islamic science.[24]

I would like to contest these views. First, I think that the elements taken by Copernicus from Arabic astronomy guaranteed him a retention of a *partial continuity* with the science of his predecessors, despite the fact that he rejected the idea of geocentrism, commonly assumed in medieval times both in Aristotelian and in Ptolemaic theories of the world. In other words, he accepted as much as possible from the previous science, which enabled him to construct in the *De revolutionibus* a more general astronomical theory than has ever been formulated before.

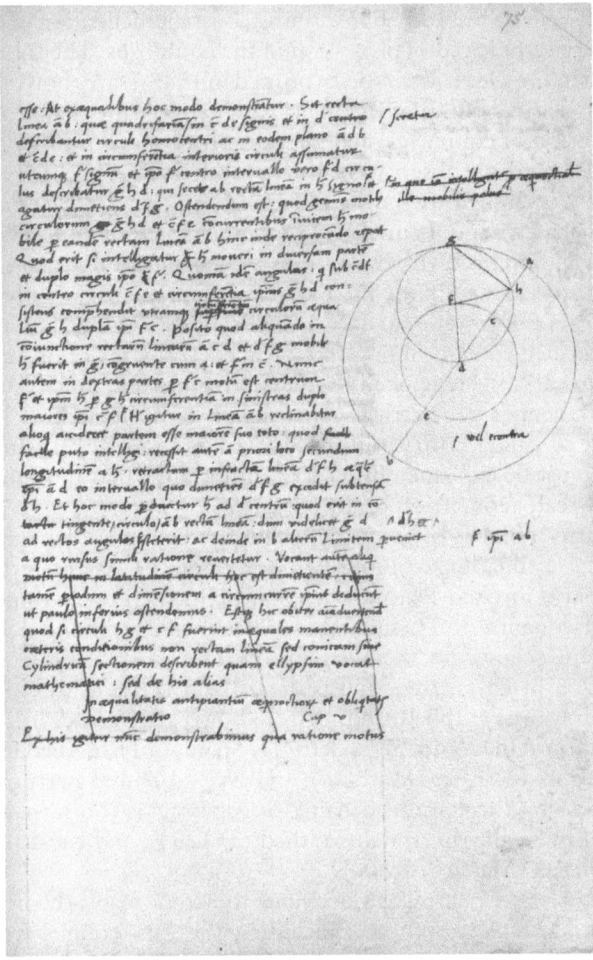

Figure 3.3   Copernicus's "Tusi couple" from the autograph of the *De revolutionibus*, Cracow, Jagiellonian Library.

But this does not impugn his originality. Moreover, such a view does not require us to deny the achievements of Arabic astronomy either, including its real originality. This is what I shall now proceed to demonstrate.

Copernicus's generalized correspondence principle, scientific method, and "dwarfs on the shoulders of giants"

It is my thesis that the most fundamental methodological core of the astronomical theories postulated by Copernicus is the idea that these theories are linked with older astronomical theories by certain generalized principles of correspondence. [25] This is realized in an analogical manner similar to the way a whole group of revolutionary theories, including, for example, the special theory of relativity, the general theory of relativity, and quantum mechanics link to the earlier tradition of classical physics. I will illustrate this claim by taking one example from the *De revolutionibus*.[26]

According to Copernicus's theory given in *De revolutionibus*, book III, chapters 3 and 6, the function of precession (that is the algebraic equivalent of the geometric-arithmetical model of precession assumed originally by him) is composed of two periodically changing parts: the first, called "the mean precession," is uniform, and the second, called "the trepidation or libration of precession," is nonuniform. The former is a linear changing function with a period of 25,816 Egyptian years (of 365 days), the latter is sinusoidal function with a period of 17,171 Egyptian years. The resultant Copernicus's precession, denoted as $P_C(t)$, counted conventionally from the observations of the star γ–Arietis (the first zodiacal star listed in the star catalog of the *Almagest*) has the following form:

$$P_C(t) = A_C + B_C t - C_C \sin 2\, \varphi(t) \tag{1a}$$

$$\varphi(t) = D_C + E_C t, \tag{1b}$$

where:

t—time elapsed since the birth of Christ; time unit is 1^{ey} (1 Egyptian year of 365 days),

$\varphi(t)$—function known as "inequality"; $2\,\varphi(t)$—function known as "the inequality of equinoctial points," or "double inequality,"

A_C—initial constant, the choice of values of magnitudes A_C and D_C causes the functions of precession to be relevant to the star γ–Arietis (the first zodiacal star listed in the star catalog of the *Almagest*),

B_C—angular velocity of the uniform component of precession,

C_C—amplitude of the nonuniform component of precession,

D_C—epoch position for the meridian of Frauenburg at midnight December 31, January 1 AD,

E_C—angular velocity of the inequality,

$$A_C = 5;32 = 5,53(3) \ [°], \tag{1c}$$

$$B_C = 360/25816 \ [°/ey], \tag{1d}$$

$$C_C = 1;10 = 1,16(6) \ [°], \tag{1e}$$

$$D_C = 6;45 = 6,75 \ [°], \tag{1f}$$

$$E_C = 360/3434 \ [°/ey].^{27} \tag{1g}$$

The analogous function according to Ptolemy's theory takes the following form:

$$P_{Pt}(t) = A_{Pt} + B_{Pt}\ t, \tag{2}$$

where:

A_{Pt}—initial constant, the choice of values of magnitudes A_{Pt} and B_{Pt} causes the functions of precession to be relevant to the star γ–Arietis,

B_{Pt}—angular velocity of the uniform component of precession,

$$A_{Pt} \approx 5,2780 \approx 5;1641 \approx 5;17\ [^\circ], \tag{2a}$$
$$B_{Pt} \approx 0,0100 \approx 1/100\ [^\circ/ey].^{28} \tag{2b}$$

Figure 3.4 shows graphs of functions of precession according to both theories, "C" standing for Copernicus's and "P" for Ptolemy's.[29]

The detailed mathematical comparison of the figure 3.4 with the figure 3.1 leads us to the conclusion that Copernicus's model of precessions explained in the *De revolutionibus* is linked by a certain generalized correspondence principle with the corresponding Ptolemy's model described in the *Almagest* (and Copernicus's earlier model sketched in the *Commentariolus*) analogous to the way relativistic and Newtonian mass correspond to each other.

We can show that this methodological strategy is assumed by Copernicus throughout the whole of his two astronomical theories described in the *Commentariolus* and *De revolutionibus*. In other words, in the formulation of his theories, he applied the hypothetico-deductive method of correspondence-oriented thinking in a systematic manner. And he does so, long before twentieth-century physicists, such as Max Planck, Albert Einstein, Erwin Schrodinger, and Werner Heisenberg adopted the same strategy.

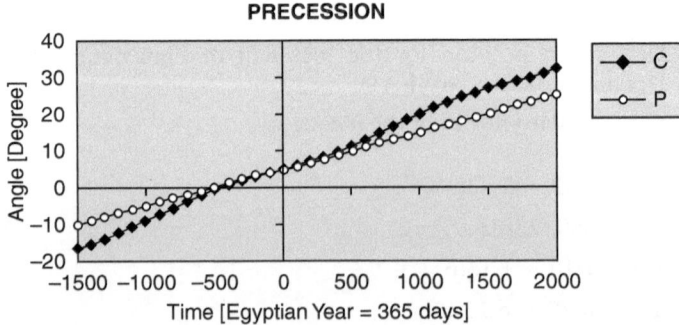

Figure 3.4   Functions of precession in Copernican and Ptolemaic theories.

Furthermore, it is necessary to emphasize that Copernicus was neither a precursor nor originator of the hypothetico-deductive method of correspondence-oriented thinking. Indeed, it was a common strategy assumed by the best scholars of all mature exact sciences (including optics, statics, and astronomy) from at least Hellenistic times, and a method applied by Archimedes, Ptolemy, and the astronomers associated with the Maragha school.[30]

WHAT, THEREFORE, IS NEW IN COPERNICUS'S THOUGHT?

It is now possible to understand why Copernicus has to be seen as an original thinker of the highest order. He formulated a harmonized mathematico-physico-astronomical theory—a *system* based on the cosmology of a mobile Earth. The theory he constructed is more general than Ptolemy's, and is free from Ptolemy's equant (which contradicts the ancient axiom of astronomy laid down by Plato's principle). Copernicus's theory was also designed to save astronomical data from the whole history of observations. In other words, it was a theory constructed to describe the history of the universe after its creation by God. This contradicts the oft-held assumption of many contemporary cosmologists that the group of theories based on the general theory of relativity were the first to offer a general history of the universe. Unfortunately, the empirical data used by Copernicus were not sufficiently precise to formulate long-period models that could save astronomical phenomena, as later noted by Tycho Brahe.[31] Nevertheless, Copernicus's astronomico-physical system, based on Buridanian physics of impetus, determined the further development of modern astronomy and physics, including the Keplerian heliocentric system, Galileo's kinematics and astronomical views, and Newtonian mechanics with its theory of gravitation.[32]

END NOTES

1. See Barker and Ariew (1991b), Hugonnard-Roche (1997), and Saliba (1984), (1987), (1995). See also Neugebauer (1968), (1975), Swerdlow (1973), (1980), Westman (1975), (1980), (1986), Swerdlow and Neugebauer (1984), Cohen (1985), and Chichester (2009).
2. See also Kokowski (1996–2009).
3. However, it is worth noticing that there are many different interpretations of T. S. Kuhn's and Feyerabend's philosophy of science, caused by the ambiguous nature of their views. For a good survey of these interpretations in the case of Kuhn's philosophy, see Jodkowski (1990), and in the case of Feyerabend's, see Oberheim (2006).
4. We can say that such cases are the ideal types in the Weberian sense. See Islam (1988).

5. See Kokowski (2004, 61).
6. See also Kokowski (2008b).
7. The problem of translation and transformation of the old theories to new ones was particularly brought to prominence in the works of T. S. Kuhn and P. K. Feyerabend. See Kuhn (1962), Feyerabend (1979).
8. It is Michael Heller's expression—see Heller (2003, 43).
9. For more details, see Kokowski (1996), (2004).
10. Reported by John of Salisbury, *Metalogicon,* iii, iv. Quoted by Sarton (1927–1947), vol. 2, part 1, 196, and Prioreschi (2002, 36). The expression, "dwarfs standing upon the shoulders of giants," was often cited in the medieval and modern times. On the history of the expression see Guyer (1930), Sarton (1935), Ockenden (1936), Klibansky (1936), de Ghellinck (1945), Merton (1965), Jeauneau (1967), and Prioreschi (2002). Recently, it was also applied by Hawking (2003).
11. See Eliade (1993), (1998).
12. "Praise be to God Who brings forth good and Who inspires truth...We wish to present a summary of astronomy as a memento for one of our dear friends, and we ask God to grant us success for its completion; for He is the One who grants success and to Him is the Final Return." (al-Tūsī 1993, 90).
13. Among the many various literary and artistic pursuits which invigorate men's minds, the strongest affection and utmost zeal should, I think, promote the studies concerned with the most beautiful objects, most deserving to be known. This is the nature of the discipline which deals with the universe's divine revolutions, the asters' motions, sizes, distances, risings and settings, as well as the causes of the other phenomena in the sky, and which, in short, explains its whole appearance. What indeed is more beautiful than heaven, which of course contains all things of beauty? This is proclaimed by its very names (in Latin), *caelum* and *mundus,* the latter denoting purity and ornament, the former a carving. On account of heaven's transcendent perfection most philosophers have called it a visible god. If then the value of the arts is judged by the subject matter which they treat, that art will be by far the foremost which is labeled astronomy by some, astrology by others, but by many of the ancients, the consummation of mathematics. Unquestionably the summit of the liberal arts and most worthy of a free man, it is supported by almost all the branches of mathematics. Arithmetic, geometry, optics, surveying, mechanics and whatever others there are all contribute to it.

Although all the good arts serve to draw man's mind away from vices and lead it toward better things, this function can be more fully performed by this art, which also provides extraordinary intellectual pleasure. For when a man is occupied with things which he sees established in the finest order and directed by divine management, will not the unremitting contemplation of them and a certain familiarity with them stimulate him to the best and to admiration for the Maker of everything, in whom are all happiness and every good? For would not the godly Psalmist (92:4) in vain declare that he was made glad through the work of the Lord and rejoiced in the works of His hands, were we not drawn to the contemplation of the highest good by this means, as though by a chariot?" (Copernicus 1978, 7)

14. See Neugebauer (1975, 3–5).
15. It is also worth mentioning that Copernicus not only collected and used the results of the astronomical observations made by other astronomers in constructing his own astronomical system, but he also made his own astronomical observations, at least from 1497 to 1541.
16. See Neugebauer (1975 7–13), Rashed, ed. (1997).
17. See Guessoum (2008) and Kokowski (2009).
18. Several authors think that a heliocentric system was built long before Copernicus (see e.g., Russo (2004, 78–85). Such authors make a serious mistake. First, they do not differentiate between the idea of *heliocentrism* (the physical Sun is at the center of the universe) and the idea of a *heliocentric system* (physico-mathematical theory with the idea that the physical Sun is at the center of the universe). If we differentiate between these two ideas, we can see that the idea of heliocentrism is of ancient origin, since it stems from ancient India and Greece, but the idea of a heliocentric system is the unique result of Kepler's achievement in the seventeenth century. By contrast, in Copernicus's versions of the astronomical system, described first in the *Commentariolus* and then in the *De revolutionibus*, it is not the real, physical Sun (as in Kepler's system), but only the abstract point, the so-called middle Sun (that is the center of orbit of the Earth) that is at the center of the universe.
19. A description of this mechanism is given by al-Tūsī (1993, 429–437) (edited, commented, and translated by F. J. Ragep); Kennedy and Roberts (1959); Neugebauer (1968); Swerdlow and Neugebauer (1984); and Bono (1995).
20. Already Gingerich (1986, 74) noticed the great importance of this point: Nevertheless, the whole idea of criticizing Ptolemy and eliminating the equant is part of the climate of opinion inherited by the Latin West from Islam. The Islamic astronomers would probably have been astonished and even horrified by the revolution started by Copernicus. Yet his motives were not completely different from theirs. In eliminating the equant, and even in placing the planets in orbit around the sun, Copernicus was in part trying to formulate a mechanically functional system, one that offered not only a mathematical representation but also a physical explanation of planetary motions. In a profound sense he was simply working out the implications of an astronomy founded by Ptolemy but transformed by the Islamic astronomers. Today that heritage belongs to the entire world of science.
21. See, for example, Duhem (1913–1959) and vol. 2 (1913, 487).
22. In his *Al-Shukūk 'alā Batlamyūs* (*Doubts Concerning Ptolemy*), Alhazen states:
    Ptolemy assumed an arrangement (hay'a) that cannot exist, and the fact that this arrangement produces in his imagination the motions that belong to the planets does not free him from the error he committed in his assumed arrangement, for the existing motions of the planets cannot be the result of an arrangement that is impossible to exist... [F]or a man to imagine a circle in the heavens, and to imagine the

planet moving in it does not bring about the planet's motion." (quoted in Sabra 1978, 121, n. 13).

23. Such scholars include Neugebauer (1968), Swerdlow (1973), Swerdlow and Neugebauer (1984), Cohen (1985), and Barker and Ariew (1991a, 1991b). For more details see Kokowski (2004, 21–30).

24. Such scholars include Nasr (1987, 174), Fernini (1998, 381), and Boutamina (2006, 383–385). For more details see Guessoum (2008).

25. Before my studies, only Moesgaard (1974a, 91) noticed that "a correspondence which includes phenomenological equivalence for ancient times only" better describes relationships between these theories (i.e., Copernicus's and Ptolemy's theories) on the level of saving phenomena opposed to simple "equivalence" of models of these two theories. This right idea was overlooked by other historians of mathematical astronomy, including Swerdlow (1980) (though, on pages 217–218, he noticed "that Copernicus's precession corresponds very closely to Ptolemy's from −300 to + 200, covering the period of the observation used by Ptolemy"), Swerdlow and Neugebauer (1984).

26. I analyzed this example in Kokowski (1996), (2004). You can find more details there.

27. Cf. *De revolutionibus*, book III, chapters 1–6. For a more precise description of the function precession, see Dobrzycki (1965), esp. pp. 32–39, Swerdlow (1980), esp. pp. 213–219, Swerdlow and Neugebauer (1984, 134–148), and Kokowski (1996, 51–70).

28. Cf. *Almagest*, book I, chapter 8, and book VII, chapter 2. The given values of constants are calculated from observational data given in the *Almagest*, book VII, chapter 2 and *De revolutionibus*, book III, chapter 2.

29. See Kokowski (2004, 65).

30. See Kokowski (1996), (2000), (2004).

31. See Blair (1990), Evans (1998, 282–283), Kokowski (2004, 23, 90, 106, 110).

32. See Kokowski (1996), (2004).

BIBLIOGRAPHY

Bala, Arun. 2006. *The Dialogue of Civilizations in the Birth of Modern Science.* New York: Palgrave Macmillan.

Barker, Peter, and Roger Ariew, eds. 1991a. "Studies in Philosophy and the History of Philosophy." *Revolution and Continuity: Essays in the History and Philosophy of Early Modern Science,* vol. 24. Washington, DC: The Catholic University of America Press.

———. 1991b. "Introduction." In *Revolution and Continuity*, edited by P. Barker and R. Ariew, pp. 1–19. Washington, DC: The Catholic University of America Press.

Blair, Ann. 1990. "Tycho Brahe's Critique of Copernicus and the Copernican System." *Journal of the History of Ideas* 51: 355–377.

Bono, Mario di. 1995. "Copernicus, Amico, Fracastoro and Tusi's Device: Observations on the Use and Transmission of a Model." *Journal for the History of Astronomy* 26: 133–154.

Boutamina, Nasr E. 2006. *L'Islam fondateur de la Science: la Renaissance et les Lumieres les siecles de plagiat.* Paris: Al-Bouraq.

Chichester, Friedel Weinert. 2009. *Copernicus, Darwin & Freud: Revolution in the History and Philosophy of Science.* Malden, MA: Wiley-Blackwell.

Cohen, I. Bernard. 1985. *Revolution in Science.* Cambridge, MA, and London: The Belknap Press of Harvard University Press.

Copernicus, Nicholas. 1978. *Complete Works.* Vol. 2. *On the Revolutions.* Edited by Jerzy Dobrzycki. Translation and commentary by Edward Rosen. Warsaw–Cracow: Polish Scientific Publishers.

———. 1985. *Complete Works.* Vol. 3. *Minor Works.* Edited by Paweł Czartoryski. Translation and commentary by Edward Rosen with the assistance of Erna Hilfstein. Warsaw–Cracow: Polish Scientific Publishers.

Dobrzycki, Jerzy. 1965. "Teoria precesji w astronomii średniowiecznej." *Studia i Materiały z Dziejów Nauki Polskiej,* Seria C, z. 11: 3–47.

———, and Richard L. Kramer. 1996. "Peuerbach and Maragha Astronomy? The Ephemerides of Johannes Angelus and Their Implications." *Journal for the History of Astronomy* 27: 187–237.

Duhem, Pierre. 1913–1959. *Le système du monde. Histoire des doctrines cosmologiques de Platon à Copernic* 1–10. Paris: Hermann.

Eliade, Micrea. 1998. *Mit wiecznego powrotu.* Warszawa: Wydawnictwo KR.

———. 1993. *Traktat o historii religii.* Translated by Jan Wierusz-Kowalski. Łódź: Wydawnictwo Opus.

Evans, James. 1998. *The History and Practice of Ancient Astronomy.* New York: Oxford University Press.

Fernini, Ilias. 1998. *A Bibliography of Scholars in Medieval Islam.* Abu Dhabi: Cultural Foundation.

Feyerabend, Paul K. *Philosophical Papers*, vol. 1: *Realism, Rationalism and Scientific Method* (1981), vol. 2: *Problems of Empiricism* (1981), vol. 3: *Knowledge, Science and Relativism* (1999). Cambridge, New York, and Melbourne: Cambridge University Press.

———. 1979. *Jak być dobrym empirystą (How to Be a Good Empiricist).* Translated by Krystyna Zamiara. Warszawa: Państwowe Wydawnictwo Naukowe (This book is a translation of selected Feyerabend's papers from 1962 to 1970).

———. 1975/1988. *Against Method. Outline of an Anarchistic Theory of Knowledge,* 2nd modified ed. London: NLB Ltd.

Ghellinck J., de. 1945"Nani et gigantes." *Buletin du Cange. Archivum latinitatis medii aevi. 19e et 20e années: 1943–1944* 18: 25–29.

Gingerich, Owen. 1986. "Islamic Astronomy." *Scientific American* 254: 74.

Guessoum, Nidhal. 2008. "Copernicus and Ibn al-Shatir: Does the Copernican Revolution Have Islamic Roots?" *The Observatory* 128: 231–239.

Guyer, F. E. 1930. "The Dwarf on the Giant's Shoulders." *Modern Languages Notes* 45: 398–402.

Hawking, Stephen. 2003. *On the Shoulders of Giants: The Great Works of Physics and Astronomy.* Philadelphia, PA: Running Press Book Publishers.

Heller, Michael. 2003, *Creative Tension: Essays on Science and Religion.* Philadelphia, PA, and London: Templeton Foundation Press,

Hugonnard-Roche Henri. 1997. "Influence de l'astronomie arabe en Occident medieval." In *Histoire des sciences arabes,* vol. 1, edited by R. Rashed, pp. 309–328. Paris: Editions du Seuil.

Islam, Nazrul. 1988. "From Ideal Type to Pure Type: Weber's Transition from History to Sociology." In *Karl Marx and Max Weber: Perspectives on*

Theory and Domination, edited by N. Islam, B. K. Jahangir, and S. I. Khan, pp. 93–110. Dhaka: Center for Advanced Researches in Social Sciences, University of Dhaka.

Jeauneau, Edouard. 1967. "Nani gigantum humeris insidentes': Essai d'interpretation de Bernard de Chartres." *Vivarium* 5: 79–99.

Jodkowski, Kazimierz. 1990. "Wspólnoty uczonych, paradygmaty rewolucje naukowe." *Realizm—Racjonalność—Relatywizm* 22. Lublin: Wydawnictwo Uniwersytetu im. Marii Curie-Skłodowskiej.

Kennedy, Edward S., and Victor Roberts. 1959. "The Planetary Theory of Ibn al-Shatir." *Isis* 50: 227–235.

Klibansky, Raymond. 1936. "Answer to Query n° 53—Standing on the Shoulders of Giants." *Isis* 26: 147–149.

Kokowski, Michal. 2009. *Różne oblicza Mikołaja Kopernika. Spotkania z historią interpretacji* (*Different Faces of Nicholas Copernicus: Meetings with the History of Interpretation*). Kraków–Warszawa: Instytut Historii Nauki PAN, Polska Akademia Umiejętności.

——. 2008a "Kulturowy charakter wielkich zmian naukowych" ("The Cultural Character of Great Scientific Changes.") *Zagadnienia Naukoznawstwa* 44, no, 1: 83–102.

——. 2008b. "The Problem of Continuity and Discontinuity in the Development of Science from Ancient Times to the Present: a Reappraisal." In *Styles of Thinking in Science and Technology: Proceedings of the 3rd International Conference of the European Society for the History of Science. Vienna, September 10–12, 2008*, edited by Hermann Hunger, Felicitas Seebacher, and Gerhard Holzer, pp. 227–254. Vienna: Austrian Academy of Sciences Press.

——, ed. 2007. *The Global and the Local: The History of Science and the Cultural Integration of Science. Proceedings of the 2nd International Conference of the European Society for the History of Science (Cracow, September, 6–9, 2006)* [E-book (with the online version)]; Cracow: The Press of the Polish Academy of Arts and Sciences. http://www.2iceshs.cyfronet.pl/proceedings.html; CD-ROM edition 2008.

——. 2006. "Nicholas Copernicus in Focus of Interdisciplinary Research." [Paper given during the 2nd International Conference of the European Society for the History of Science (Cracow, September 6–9, 2006)] in *Organon* 35: 73–84; and in Kokowski ed. 2007a, 2008a, pp. 333–341.

——. 2004. *Copernicus's Originality: Towards Integration of Contemporary Copernican Studies*. Warsaw-Cracow: Wydawnictwa IHN PAN.

——. 2001. *"Thomas S. Kuhn (1922–1996) a zagadnienie rewolucji kopernikowskiej (Thomas S. Kuhn (1922–1996) and the Issue of the Copernican Revolution)."* In *Studia Copernicana*, vol. 39. Warszawa: Wydawnictwa IHN PAN.

——. 2000. "Historia epicykliczno-deferencjalnego modelu ruchu Księżyca i hipotetyczno-dedukcyjna metoda myślenia korespondencyjnego" ("The History of the Epicycle-Deferent Model of Lunar Motion and the Hypotetico-Deductivie Method of Cor-Respondence-Oriented Thinking.") *Kwartalnik Historii Nauki i Techniki* 3–4: 77–108.

——. 1996. "Copernicus and the Hypothetico-Deductive Method of Correspondence Thinking: An Introduction." *Theoria et Historia Scientiarum*: 7–101.

Kuhn, Thomas Samuel. 1985. *The Copernican Revolution: Planetary Astronomy in the Development of Western Thought.* New York: MJF Books.
———. 1962/1970. *The Structure of Scientific Revolution.* Chicago, IL: The University of Chicago Press.
———. 1957. *The Copernican Revolution: Planetary Astronomy in the Development of Western Thought,* 7th ed. Cambridge, MA: Harvard University Press.
Merton Robert K. 1965. *On the Shoulders of Giants: A Shandean Postscript.* Foreword by Umberto Eco. New York: Free Press.
Moesgaard, Kristian Peder. 1974a. "Success and Failure in Copernicus' Planetary Theories." Part I, *Archives Internationales d'Histoire des Sciences* 24, no. 94: 73–111.
———. 1974b. "Success and Failure in Copernicus' Planetary Theories." Part II, *Archives Internationales d'Histoire des Sciences* 24, no. 95: 243–318.
Nasr, Seyyed Hossein. 1987, *Science and Civilization in Islam,* 2nd ed. Cambridge, MA: Harvard University Press.
———. 1958/1993. *An Introduction to Islamic Cosmological Doctrines: Conceptions of Nature and Methods Used for Its Study by the Ikhwan Al-Safā', Al-Bīrūnī, and Ibn Sīnā.* Cambridge, MA: Harvard University Press. 2nd revised ed. New York: State University of New York Press, 1993.
Neugebauer, Otto. 1975. "A History of Ancient Mathematical Astronomy." In *Studies in the History of Mathematics and Physical Sciences,* vol. 1. Berlin, Heidelberg, and New York: Springer-Verlag.
———. 1968. "On the Planetary Theories of Copernicus." *Vistas in Astronomy* 10: 89–103.
Oberheim, Eric. 2006. *Feyerabend's Philosophy.* Berlin and New York: Walter de Gruyter.
Ockenden, R. E. 1936. "Answer to Query n° 53—Standing on the Shoulders of Giants." *Isis* 25, no. 2: 451–452.
Prioreschi, Plinio. 2002. "The Idea of Scientific Progress in Antiquity and in the Middle Ages." *Vesalius* 8, no. 1: 34–45.
Ptolemy, Claudius. 1952. "Almagest" In *The Great Books of the Western World.* Translated by R. Catesby Taliaferro. Chicago, IL, London. Toronto, and Geneva: Encyclopaedia Britannica, Inc.
Puerbach, Georg, and Joannes Regiomontanus. 1496. *Epitome Almagestum.* Venetiae
Ragep, F. Jamil. 2001. "Tūsī and Copernicus: The Earth's Motion in Context." *Science in Context* 14, no. 1–2: 145–163.
———. 1993. "Nasīr al-dīn al- Tūsī *Memoir on Astronomy (al-Tadhkira fi 'ilm al-hay'a).*" Vol. 1 *Edition, Translation, Commentary and Introduction,* Vol. 2 *Commentary and Apparatus.* "Sources in the History of Mathematics and Physical Sciences." Vol. 12 (2 parts). New York: Springer-Verlag.
Rashed, R., ed. 1997. (avec la collaboration de Régis Morelon). *Histoire des sciences arabes,* vol. 1–3. Paris: Le Seuil.
Russo, Lucio. 2004. *The Forgotten Revolution: How Science was Born in 300 BC and Why it Had to Be Reborn.* Translated by Silvio Levy. Berlin: Springer.
Sabra, A. I. 1978. "An Eleventh-Century Refutation of Ptolemy's Planetary Theory." *Science and History: Studies in Honor of Edward Rosen, Studia Copernicana* 16: 117–131.

Saliba, George. 1995. *A History of Arabic Astronomy: Planetary Theories during the Golden Age of Islam*. New York: New York University Press.

———. 1987. "The Rôle of Marāgha in the Development of Islamic Astronomy: A Scientific Revolution Before the Renaissance." *Revue de synthèse* 4: 361–373.

———. 1984. "Arabic Astronomy and Copernicus." *Zeitschrift fur Geschichte der Arabisch-Islamisch Wissenschaften* 1: 73–87.

———. 1979. "The First Non-Ptolemaic Astronomy at the Maraghah School." *Isis* 70, no. 4: 571–576.

Sarton, George. 1935. "Query n° 53: Standing on the Shoulders of Giants." *Isis* 24, no. 1: 107–109.

———. 1927–1947. *Introduction to the History of Science*, vol. 1–3. Baltimore, MD: Williams and Wilkins Co.

Swerdlow, Noel M. 1980. "Long-Period Motions of the Earth in «De revolutionibus»." *Centaurus* 24: 212–245.

———. 1973. "The Derivation and the First Draft of Copernicus's Planetary Theory. A Translation of the Commentariolus with Commentary." *Proceedings of the American Philosophical Society* 117: 423–512.

———, and Otto Neugebauer. 1984. "Studies in the History of Mathematics and Physical Sciences." *Mathematical Astronomy in Copernicus's «De revolutionibus»*, vol. 10, 2 parts. New York: Springer-Verlag Inc.

al-Tūsī, Nasīr al-dīn. 1993. *Memoir on Astronomy (al-Tadhkira fi `ilm al-hay' a)*." Vol. 1 *Edition, Translation, Commentary and Introduction*, vol. 2 *Commentary and Apparatus*. New York: Springer-Verlag.

Westman Robert S. 1986. "The Copernicans and the Churches." In *God and Nature: Historical Essays on the Encounter between Christianity and Science*, edited by David C. Lindberg and Ronald L. Numbers, pp. 76–113. Berkeley, Los Angeles, and London: University of California Press.

———. 1980. "The Astronomer's Role in the Sixteenth Century: A Preliminary Study." *History of Science* 17: 105–147.

———. 1975. "The Wittenburg Interpretation of the Copernican Theory." In *The Nature of Scientific Discovery. A Symposium Commemorating the 500th Anniversary of the Birth of Nicolaus Copernicus*, edited by Owen Gingerich, pp. 393–429. City of Washington: Smithsonian Institution Press.

4

TIME ATOMISM AND ASH'ARITE ORIGINS FOR CARTESIAN OCCASIONALISM REVISITED

Richard T. W. Arthur

INTRODUCTION

In gauging the contributions of Asian thinkers to the making of modern "Western" philosophy and science, one often encounters the difficulty of establishing a direct influence. Arun Bala and George Gheverghese Joseph have termed this "the transmission problem."[1] One can establish a precedence, as well as a strong probability that an influence occurred, without being able to find concrete evidence for it. In the face of this difficulty (which appears to occur quite generally in the history of thought), I suggest here that the influence of earlier thinkers does not always occur through one person reading others' works and becoming persuaded by their arguments, but by people in given epistemic situations being constrained by certain historically and socially conditioned trends of thought—for which constraining and conditioned trends of thought I coin the term "epistemic vectors"—and opportunistically availing themselves of kindred views from other traditions.

As a case in point, I will examine here the claim that the doctrine of Occasionalism arose in the seventeenth-century Europe as a result of an influence from Islamic theology. In particular, the Ash'arites school of *kalâm* presented occasionalism as a corollary of *time atomism*, and since to many scholars the seventeenth-century occasionalism of Cartesian thinkers, such as De la Forge and Cordemoy[2] has appeared as a direct corollary of the atomism of time attributed to Descartes in his *Meditations*; Ash'arites time atomism is often cited as the likely source of Cartesian Occasionalism. It has also been suggested that this idea of time atoms is foreign to Greek thought, and accordingly may well be an indication of influence on Arab thinkers in turn by the Buddhist sect of the Sautrāntikas of Northern India of the second or first century BCE.

These claims are directly relevant to the focus of this chapter because of the centrality of Cartesianism to the scientific revolution of the seventeenth century. In particular, Descartes's way of conceiving the world as constituted by its instantaneous state, reducing forces to the tendencies to move possessed by bodies at each instant, is arguably one of the crucial moments in the establishment of modern physics, since in its further development in the hands of Newton and Leibniz it paves the way for the concept of "instantaneous velocity" and "the creation of the calculus." The idea that these moments are discrete time atoms, moreover, and not pointlike modalities in the continuum, is of great topical relevance: time atomism is very much a live issue in the context of modern theories of Quantum Gravity, and has been suggested both in the context of the String Theoretic and Loop Quantum Gravity approaches.[3]

For various reasons, though, both these historical claims of influence—that of the Sautrāntikas on the Mutakellimim (also called Mûtakallimûn, the *kalâm* theologians), and the influence of *kalam* doctrines on Descartes—are quite contentious, most particularly because of the lack of the type of hard corroborating evidence required to establish a direct influence. I shall take a more nuanced stand, and try to articulate the epistemic vectors that may nevertheless have led to some of these apparent influences of Asian thought on ideas at the epicenter of the scientific revolution in the seventeenth-century Europe.

KALAM, THE SAUTRĀNTIKAS, AND TIME ATOMISM

Inspired by various passages in the *Koran* that seem to reserve almost all power for God alone, the Islamic theologian al-Ash'arî (873–935 CE) and his followers claimed that there is no real causation in the world save God's. Thus, strictly speaking al-Ash'arî never caused his pen to write those austere words: rather, God created al-Ash'arî's will to write them and his power to move his pen, and simultaneously caused the motion of his hand and pen, the flowing of the ink, and the absorption of the ink into the paper.[4] Neither al-Ash'arî himself, nor his hand, nor the pen can thus be said to *act*, nor to cause any further effects in the other things. The presence of the idea in al-Ash'arî's mind simply provided God with the *occasion* to simultaneously bring about the moving of his hand and quill, and the other concomitant effects. In the same way, it is not the dye that causes the cloth to become black: all causal power is God's alone.

This is the doctrine of *Occasionalism*, a subterranean doctrine that has resurfaced many times in the history of thought, though not always under that name, and certainly not always in the knowledge that it had been proposed before. The classic statement of it was given in the seventeenth century by Nicolas Malebranche: "[T]here is only one true cause because there is only one true God;...the nature or power of each thing is nothing but the will of God;...all natural causes are not *true* causes but only the *occasional* causes of natural effects."[5] Now the Ash'arites presented it as a corollary of the *atomism of time*: as Sorabji reports it:

> The Ash'arites held that every time-atom God creates an entirely new set of accidental properties, although they may be accidents of the same kind as before. If he omits to create new accidents, the substance which bore them will cease to exist. This shows why the blackness which we think is introduced into the cloth by the dye must in fact be created and re-created every time-atom by God.[6]

The universe, according to the Ash'arites, is comprised of an indeterminate number of indivisible, homogeneous particles, themselves unextended and devoid of magnitude (*kam*). When these atoms are combined, bodies result. Opinions differed as to how many atoms this would take: according to Ash'arî, some of the atomists preceding him had held that a body (or possibly two bodies) arise from the combining of two atoms. "Abu l-Hudhayl, however, argued that a body required a minimum of six atoms, corresponding to the six planes of a solid."[7] But a body does not have its accidents—color, taste, contact, motion, rest and, the like— as a whole; rather they are possessed by its constituent atoms individually. Moreover, these atoms are not everlasting, like those of the Greeks, but last no longer than the accidents adhering in them. Most of the Mûtakallimûn, reports Fakhry, held that "the atom cannot endure for two instants of time" (Proposition 6 of Maimonides's account), since duration (*baqâ*) itself can no more exist beyond an instant than can any other accident.[8] Fakhry cites Al-Bâqilânî (d. 1013) as defining an accident as "'that which cannot endure...but perishes in the second instant of its coming-to-be'—a definition for which he finds a scriptural basis in the Koran which speaks of the 'transient things' of this world (*a'rad*)."[9] Thus when a garment has been dyed red, God recreates its constituent atoms with the accident red in each successive time atom of the garment's subsequent existence. But "there would be nothing to prevent Him from creating in it the accident yellow or the accident black." He does not do so, not because this is impossible, but simply because

> God has decreed as a matter of habit that the succession of accidents shall correspond to a certain pattern; so that the colour black, e. g., shall not appear in the garment unless it is brought into contact with a black dye, [and] that it shall [not] be followed, upon its instantaneous cessation, by any save the colour black. But it is clear that God, who is the ultimate agent, could alter this course of habit freely.[10]

Now this view is extremely reminiscent of Malebranche's doctrine. And coupled with the fact that Descartes's upholding of the thesis of the equivalence of conservation with divine recreation has been widely taken to commit him to temporal atomism, this has led many to see an Islamic origin for Cartesian Occasionalism. This is well summarized by Sorabji:

> Maimonides reports further corollaries of this view: things do not have essential natures of their own, since it is God who creates all their properties. Again, things have no tendency of their own to persist—it is this which most of all would encourage a sense of precariousness. The doctrine that our continuation requires that God re-create us from moment to moment is repeated in Descartes, and from there it influenced the seventeenth-century occasionalists. The best known of them, Malebranche, agreed that all causation involves creation.[11]

But I will defer treatment of the question of Descartes and Cartesian Occasionalism to the next section. First, I want to discuss the second of the claimed influences mentioned above, concerning the origins of Islamic atomism itself. For as several authors have pointed out, the atomism of the Mutakellimim seems quite different from that of the ancient Greeks in several respects. Most scholars agree that Leucippus was motivated to introduce atoms to answer the Eleatics's devastating critiques of plurality and change: the atoms have most of the qualities of the Parmenidean One or Being, except that they are many and different, and move in the void (Non-being). Leucippus, Democritus, and Epicurus explained phenomenal qualities and their changes in terms of atomic shapes, sizes, motions, and rearrangements. As Otto Pretzl argued, the role of the atoms of the early Greek atomists was to function as the permanent substances underlying change.[12] The *kalâm* atoms, by contrast, are pointlike and lacking in extension, indeterminate in number, and simply possess phenomenal qualities like color and taste, rather than reductively explaining them. But what is most striking in the contrast is the instantaneousness of the atoms of the Mutakellimim, the fact that they do not endure longer than an instant. Far from constituting what is constant underlying phenomenal change, they are not themselves substantial, but ephemeral.

These differences have led commentators to argue that Islamic atomism is too dissimilar from the Greeks' to have been inherited from them directly. In particular, Otto Pretzl suggested that it had traveled via the Gnostics[13]; and D. B. MacDonald speculated that the idea of time atoms probably arose as a Muslim heresy in the dark centuries immediately after the death of the Prophet, and that it might be an indication of influence on the Arab thinkers from the Buddhist sect of the Sautrāntikas of the second or first century BCE.[14] The Sautrāntikas (more properly,

Saṃkrāntikas, and probably identical with the Dārṣṭāntikas) are a somewhat obscure Buddhist sect about which little is known definitively. What is known is that they were active in the first or second century BCE, that they proposed a theory of "point-atoms" or events (*dharmas*) that had a merely momentary existence.[15] This is taken up by later Indian Buddhist philosophers such as Dharmakirti and Dignaga, who also considered atoms to be point-atoms, durationless events. The brunt of this argument for Indian influence, however, is the lack of any Greek theory combining temporal and material atomism in the same way as the Ash'arites atomic theory. Similarly, in his *History of Islamic Philosophy*, Majid Fakhry writes of the Muslim atomic theory that "it is noteworthy that some of its important divergences from Greek antecedents, such as the atomic nature of time, space and accidents, the perishability of atoms and accidents, appear to reflect an Indian influence."[16]

Such an influence is by no means impossible, but it is very difficult to establish. Certainly, the Arabs were open to learning from the Indians: as Fakhry reports, "one of the earliest works to be translated into Arabic was an Indian astronomical treatise, the *Siddhanta* of Brahmagupta, which in the Arabic version of al-Fazari played an important role in the development of Islamic astronomy."[17] But he is forced to concede that, concerning "the more philosophical elements in Indian thought that might have influenced the Arabs, we are at once struck by their relative scarcity or triviality when compared to the rich stream of ideas that came from Greece."[18] Certain figures who are known to have had substantial engagement with the philosophy of the Indians, such as al-Bîrunî (d. 1048) and al-Râzî (d. ca. 925), came too late to be credited with the origination of Islamic atomism. But there is a report that the great Islamic philosopher, al-Kindî (805–873) had transcribed "an anonymous treatise on the *Religious Beliefs of the Indians*, which was in circulation among the Arabs by the end of the eighth century."[19] And by this time Indian atomic theory was well established: "The two Buddhist sects of Vaibhashika and Sautrāntika, the two Brahmin sects of Nyaya and Vaishashika, as well as the Jaina sect, had evolved by the fifth century an atomic theory, apparently independent of the Greek..."[20] As Arun Bala has argued:

> These schools were quite influential at the time the Arab-Muslims emerged as the dominant power after their conquest of the Sassanian Empire, and their views must have been discussed at the centres of learning at Jundishapur, where scholars from the Hellenic and Indian worlds met. Arabic theologians might have found such views attractive because Indian atomic views were closely linked with religious traditions such as Hinduism, Buddhism and Jainism—and did not have the association with atheism they had in the Hellenistic world....Moreover, the association

with religious thought also produced other similarities between Indian and Arabic atomism—*kalam* atomism did not deny the existence of minds independent of matter, belief in an afterlife, the need for spiritual cultivation, or the possibility of communion with a transcendent reality.[21]

All of this is undeniable. Nevertheless, the case for influence rests very heavily on the features of Islamic atomism that are alleged to have no correlates in Greek philosophy, in particular, time atomism. On this score, however, these arguments for an Indian influence on *kalâm* are seriously weakened by Richard Sorabji's detailed analysis of the many parallels between the positions of the ancient Greeks and the points in the debates between Abû l-Hudhayl (died 841), his nephew Nazzām (died c. 846), and Dirār (died 815 or earlier). For according to Sorabji, too much attention has been paid by thinkers such as Fakhry to the Presocratic atomists in the Hellenic tradition, and too little attention has been paid to later atomisms, such as that of Diodorus Cronus, to Epicurus's theory of the minimal parts within atoms, and again to the views of Xenocrates, and late Neoplatonist theories such as Damascius's theory of infinitely divisible leaps of motion. Once these views are taken into account, some vivid and detailed points of comparison can be established between the views of the earliest Islamic atomists and corresponding Greek views.

For example, Nazzām, like Diodorus and Damascius, advocated infinitely divisible leaps.[22] And Dirār can be seen to be following the Neoplatonist view that sensible individuals are bundles of properties, properties they refer to as *accidents*, and also as following Epicurus in his attack on the Anaxagorean theory of latency maintained by Nazzām.[23] More importantly for our concerns here, Sorabji finds in Diodorus "an argument that the present must be a time atom," and in Epicurus the view that there are minimal times that can be successive without touching. As he reports, Epicurus's theory is described by his follower, the atomist Demetrius, as positing "something like this": "The thing happens whenever, from the place where this one emerged, the neighbouring one will follow at once at the next (*hexes*) time, which is a minimal time." (word ordering slightly altered.)[24]

Thus according to Sorabji's analysis there are indeed Greek counterparts to the Islamic time atoms reported by Maimonides and upheld by Islamic thinkers such as Abû l-Hudhayl. Indeed, after a detailed analysis of the points of correspondence, Sorabji finds the controversy over atomism between Nazzām and l-Hudhayl "full of Greek resonances."[25] As for the fact that the Ash'arites do not use atoms to explain the physical world, this is no less true of Diodorus. And even regarding the unextended atoms of l-Hudhayl, Sorabji is able to allude to the possibility of a Greek source, with Wolfson's identification of an extant ninth- or tenth-century Arabic work that misrepresents Democritus's spherical atoms as points.[26]

Still, Sorabji does "not regard the presence of Greek influence as *excluding* Indian," and cites the example of the Indian atomist Kanade, founder of the Nyaya-Vaiseshika school, whose *floruit* is dated by some scholars as around 100 CE. Kanade, according to Sorabji, uses the same argument that is found in earlier Greek thought, namely that without atoms, the large and the small would be equally big, on account of the infinity of the parts of something that is infinitely divisible.[27] According to Sorabji this also occurs later in *kalâm* writings, but, fascinatingly, with the same illustration that Kanade gives, of the mountain and the mustard seed. Especially suggestive of a three-way influence is Kanade's claim, reported by Sorabji, that six is the minimum number of atoms needed to produce magnitude. For this occurs in Abû l-Hudhayl, who reasons "that the surrounding atoms can be arranged above and below, in front and behind, to the right and the left." [28] This argument is found in Aristotle as a correction to Democritus's view that there are just four differences of position.[29] All of this seems to indicate that much more work needs to be done on the detailed correspondences between the arguments of the various schools extant in the ninth and tenth centuries. It is by no means unlikely that the Muslim philosophers could have learnt of the Indian atomic theories, and made use of them where it suited their purposes. The earlier forms of Islamic atomism of the ninth century seem indebted to Greek ideas, although this does not belittle the Mutakellimim's own original contributions, and by no means excludes influence from Jainist, Buddhist, and Hindu sources; and the Indian atomisms themselves may well have been influenced by Greek ideas in their turn, despite the skepticism of Indian historians of philosophy.

However, we can also say that, even if time atomism was not alien to Greek philosophy, it is in certain schools in Indian philosophy where this is systematically tied to theses concerning the instantaneity of the existence of atoms and their accidents. And even though al-Râzî's study of Indian thought came too late for it to be a factor in the origination of Islamic atomism, it was not too late to have contributed to his contemporary al-Ash'arî's fusing of atomism with occasionalism. If al-Ash'arî' were familiar with the complete ephemerality of the point-atoms of the Sautrântikas, that is a factor that would have had a definite appeal in an atomist context where it is above all necessary to show how all things depend on God as cause. Even here, though, it seems to me that some caution is necessary. For this Buddhist sect's universe is fragmentary: all that exist for them are point-atoms, each becoming manifest as it is individually experienced on different occasions; but there is nothing in this universe corresponding to Aristotle's τὸ νυν, a worldwide instant that can be experienced by everyone existing at the same time. Strictly speaking, in fact, both time atomism and the idea of discrete world-states following one another are alien to the Buddhists' philosophy of dependent origination. But the idea of the world at an instant is implicit in the time atomism of both Epicurus and Diodorus, coming as they do after Aristotle. And Al-Ashʿarî's occasionalism, like the occasionalism of the later Cartesian thinkers, appeals to time-slices, instantaneous world-states, not spatially isolated momentary fragments.

Thus, if we were to make a tentative list of the moments of thought that are crucial to *kalâm*, this notion of a worldwide instant would be among them. This is an example of what I have called an "epistemic vector." A much more crucial constraint on their thinking, of course, is the fundamental Islamic tenet of the total dependence of the created world on God, something that is quite absent from the Buddhist philosophical approach. The same goes for the concomitant denial that there is, strictly speaking, any agent at all save for God. In al-Ghazâlî this becomes much more pronounced with the critique of essences or natures, and the undermining of the idea that there is a necessary relation between cause and effect. But these are somewhat later elaborations of the Ashʿarite view by al-Ghazâlî in his *Incoherence of the Philosophers (Tahaifut al-Falaisifa)*, a vigorous defense of *kalâm* against Peripatetic-based philosophy like Avicenna's (Ibn Sīnā, 980–1037), and a work that would itself gain fame as the target of Averröes's rejoinders on behalf of Aristotelianism. So if we restrict our consideration to the earlier period of *kalâm*, and to Ashʿarī's formulation of the doctrine, some of the crucial moments or epistemic vectors that issue in the doctrine could tentatively be listed as follows:

1. The absolute omnipotence of God, and the absolute dependence of creatures on God for their existence.
2. The denial that there is any creative or causal agent apart from God.
3. The supposition that the world exists in a succession of worldwide instants.
4. The supposition that everything that exists consists of atoms and their accidents.

In this context, the Mutakellimim would have had the resources to formulate their philosophy without having to go beyond the borders of Islamic theological doctrine and their criticisms of Greek views. In particular, as Sorabji points out, the identification of το νυνς (instants) with partless times is found in Diodorus Cronus, and partless times are also to be found in Epicurus.[30] But a familiarity with the views of the early Buddhists, in particular, that all atoms and their accidents are utterly ephemeral, would certainly have emboldened them in explicitly identifying instants as time atoms, leading fairly naturally to the thesis that atoms and their accidents do not endure longer than an instant.

But actually it is not necessary to resolve the question of the origin of Islamic Atomism to understand the origins of occasionalism. For it is possible that occasionalism, while certainly fortified in al-Ash'arî's philosophy by time atomism, is not necessarily consequent on it. This seems to be the case also with respect to Cartesianism. For several scholars, myself included, have contested the attribution of a time atomism to Descartes, even though it can be found clearly advocated by two of his occasionalist followers, Louis de la Forge and Gerauld Cordemoy. But the fact that time atomism is absent from the philosophy of Nicolas Malebranche, perhaps the most influential of the Cartesian Occasionalists, certainly requires us to think again about the relationship of occasionalism to time atomism. So let me turn now to the origins of Cartesian occasionalism and the question of the influence of *kalâm* on it.

AUGUSTINE AND THE ORIGINS OF CARTESIAN OCCASIONALISM

It is beyond question that for the first several hundred years of its existence what we like to call Western philosophy, was so interwoven with Islamic philosophy that the former would have been quite different without it. Certainly, there's no more question about the huge influence on medieval philosophy of the writings of Avicenna and Averröes than there is about the fact that those thinkers were responding to Aristotle and the Greeks as well as al-Ghazâlî, and were at the same time a major source of knowledge of those doctrines for the Christian West. The doctrines of the Mutakellimim were certainly known in Europe, not least on account of Maimonides's summary, and their influence on the doctrines of the likes of Gerard of Odo and Nicolas d'Autrecourt have been established in some detail.[31] So the doctrines of the Mutakellimim (either through Maimonides or refracted through other authors, such as the so-called Zenonists) would certainly have been available to Descartes, as well as to his followers, as potential influences on their own thought. But it may well be questioned whether internal evidence supports an explicit indebtedness of Descartes or Malebranche to the *kalâm* philosophy, as opposed to something more nuanced: in the case of Malebranche, a pushing of his views in the same occasionalist direction as the Ash'arites by certain doctrinal certitudes they had in common, quite independent of the question of time atomism. Majid Fakhry puts his finger on one of them, right at the beginning of his introduction to his *Islamic Occasionalism and its Critique by Averröes and Aquinas*. After acknowledging the well-known origin of Malebranche's occasionalism in his attempt to solve the inherent difficulties of Cartesian mind-body interaction (and, he might have added, body-body interaction too), he adds:

> But it is not sufficiently recognized that his notion of God's direct role in activity is directly affiliated to St. Augustine and his "theology of grace." Even a cursory perusal of Malebranche's major work *Recherche de la Vérité*

is sufficient to show the extent of St. Augustine's influence on his thought. In point of method, perhaps, the Cartesian influence predominates; but the theological inspiration, which determined the shape of his occasional-ism, is unmistakably Augustinian. As a matter of fact, Islamic occasional-ism, as we are going to see, is inspired by precisely the same Augustinian motive, namely, the vindication of the absolute omnipotence and sover-eignty of God and the utter powerlessness of the creature without Him.[32]

Indeed, substantial light can be thrown on occasionalism and its origins by reference to the thought of St. Augustine, Bishop of Hippo (354–430). There are two points to be made in this connection, as Sorabji has observed. First, Augustine is explicitly acknowledged by Malebranche as a source of inspiration for his occasionalism.[33] Second, Augustine is not himself an occasionalist, even though he restricts all creative and causal power so that it ultimately resides in God. This is because he does not deny that created souls have within them the power to act, but insists that in doing so "they do not create, but only make use of the forces supplied by God to bring forth what he has already created."[34] As we shall see, Augustine locates these forces in the *seminal reasons* in cre-ated beings, provided by God at Creation, a doctrine deriving from and synthesizing prior doctrines of Stoic and Neoplatonic origin. This device allows him to make room for secondary causes without denying that all such causes depend completely on God. Augustine does not allow that *bodies* can be agents in this sense, however, and here Sorabji identi-fies a third current of thought in addition to the doctrinal constraint common to both Christian and Islamic theology that "God is the only creator," and Augustine's seminal reasons: this is the Platonic doctrine that bodies are causally inefficacious in the sense that, although they can transmit motion by bodily collisions, they are incapable of *originating* motion—or, as Sorabji writes more colloquially: "only *souls*, not *bodies*, can originate motion; bodies merely transmit motion by banging into each other."[35]

This extremely influential doctrine, whose force resounds down through the centuries, can be identified as one of the main sources for later occasionalist ideas in Christian thought, if not also Islamic theology too. It appears to have been upheld by the great majority of seventeenth-century thinkers, including not only Descartes and the occasionalists, but More, Newton, Leibniz, Berkeley,[36] and many others.

The main conduit, of course, was Neoplatonism, where the doctrine was expanded by Plotinus to encompass not only human souls, but also the Intellect and the One, as the only true agents. Naturally, the appeal of such a doctrine to Christian theologians was considerable, and it can be found soon after the time of Plotinus (c. 205–260 CE) in the writ-ings of Augustine. For combining the idea that only God and souls can act, with the Christian doctrine that God is the only creator, Augustine argues that parents, farmers, builders, and so forth do not create, but simply bring out what God has already created. He cites a passage from St. Paul: "I planted, Apollos watered, but God gave the growth. So neither he who plants nor he who waters is anything, but only God who gives the growth."[37]

As Sorabji mentions, Augustine exploits this biblical passage to introduce his doctrine of *seminal reasons*. These are the seeds planted in matter at Creation from which all natural kinds unfold over the course of history, containing the reasons for all the subsequent characteristic behavior of each living thing. God has created all these seminal reasons in things at the very beginning: the unfolding of the natural kinds through history is God's "giving the growth" to them. Although God creates new souls, he is the one who unites them to their bodies, but there is nothing in the bodies that they interact with that they actually create. In fact, Augustine argues, the lack of creative power of created beings cuts deeper. For without God's continually sustaining created things in existence through every instant of their duration, all of nature would lapse into nonexistence:

> If the Creator's virtue were at any time to be missing from the created things which are to be governed, at once their species would go missing, and the whole of nature would collapse. For it is not like the case of a builder of houses who goes away once he has built, but whose work stands, even though he is missing and goes away. The world could not last like this for the duration of an eyeblink if God were to withdraw his governance from it.[38]

This is the doctrine that conservation of existing things is equivalent to God's continually creating them, the doctrine of continuous creation. As we shall see presently, this assumes a significant role in Cartesianism. As Sorabji observes, the comparison with the architect is repeated by Malebranche in his *Dialogues on Metaphysics [Entretiens sur la métaphysique]*. When Aristes suggests that for the world to be annihilated, it is not enough that God should no longer will it to exist, but that he would have to positively will it to cease to exist, Theodore replies:

> You are not thinking, Aristes. You are making creatures independent. You judge God and His works by the works of men, men who are provided nature and do not make it. Your house subsists although your architect is dead. This is because its foundations are solid and it has no connection with the life of the person who built it. It depends on him in no way. But the ground of our being depends essentially on the Creator. Though the arrangement of certain stones depends in a sense on man's will in consequence of the action of natural causes, the product is not so dependent.

But, as the universe is derived from nothing, it depends to such an extent on the universal Cause that, if God ceased to conserve it, it would necessarily revert to nothing.[39]

Commenting on this juxtaposition of passages, Sorabji writes that what Malebranche "adds, and what is missing so far as I know from Augustine, is the idea that continuation depends on continuous re-creation."[40] But there is no talk of *re*creation here at all: indeed, the exchange between Aristes and Theodore follows a passage in which Theodore makes it clear that the continued existence of the world is due to God's *continuous* willing that it should exist:

> God wills that there be a world. His will is all-powerful, and so the world is made. Let God no longer will that there be a world, and it is thereby annihilated. For the world certainly depends on the volitions of the Creator. If the world subsists, it is because God continues to will that the world exist. On the part of God, the conservation of creatures is simply their continued creation. I say, on the part of God who acts. For on the part of creatures, there appears to be a difference, since, in creation, they pass from nothing to being whereas, in conservation they continue to be. But, in reality, creation does not pass away because, in God, conservation and creation are one and the same volition which consequently is followed by the same effects.[41]

Here Malebranche precisely denies any alternation of creation and annihilation. This is the crucial contrast with Islamic occasionalism, where nothing endures beyond a single time atom, not even duration itself. But what Malebranche's occasionalism has in common with that of al-Ghazālī, and where they differ from Augustine, is in the denial that created things have either natures or causal powers of any sort. Al-Ghazālī (perhaps also influenced by the ideas of the Greek skeptics) argues that there is no necessary connection between the contact of a piece of cotton with fire and its burning: God could create either one without the other, but chooses not to. Malebranche agrees: the appearance of natures, and of necessary connections between cause and effect, are simply reflections of God's habits in willing things in similar ways. There are, of course, many differences as well as similarities between the views of Malebranche and al-Ghazālī, as was to be expected. But I do not want to dwell on these. My point is that the continuous creation doctrine does not presuppose time atomism; and while it is compatible with occasionalism, as in the case of Malebranche, it does not require it, as evidenced by the case of Augustine. This is of crucial importance for understanding the views of Descartes, who, as we shall see, acts as a bridge between Augustine and Malebranche. And this in turn is of the utmost relevance to the origins of modern science, to which I now turn.

DESCARTES'S INSTANTANEISM

One of the most outstanding contributions that René Descartes made to modern science was the idea of an *instantaneous state of motion*. This he had already conceived in 1629–1633 when he wrote his first treatise on physics, *Le Monde* (*The World*), but after the Church's censure of Galileo that year, he decided that it would be unwise to risk publishing a treatise so explicitly Copernican. So he contented himself with publishing essays on Optics, Meteorology, and Geometry instead, and his ideas about the physics of motion made their public debut somewhat later in his *Principles of Philosophy* of 1644. In *The World*, Descartes had appealed to the principle that "each individual part of matter continues always to be in the same state so long as collision with others does not force it to change that state"—a relatively uncontroversial conservative principle—but then insisted on regarding *motion* as a state. This is ostensibly paradoxical: a state is something that does not change, whereas a motion is the paradigm example of a change. Nevertheless, Descartes persisted, arguing that the school philosophers' ideas of motion (motion with respect to form, motion with respect to heat, and motion with respect to quantity) were far more obscure than the geometrical notion that he was advocating. In the *Principles*, this argument is abandoned in favor of the following terse formulation:

> *The first law of nature: each and every thing, in so far as it can, always continues in the same state; and thus what is once in motion will always continue to move.*[42]

This is recognizably similar to Newton's famous First Law of Motion, which should not be too surprising since, as modern scholarship has established, Descartes's laws as stated in the *Principles* were indeed the source for Newton's own. The correspondence with Newton's Law of Inertia is even clearer when Descartes's first law is supplemented with his second:

> *The second law of nature: all motion is in itself rectilinear; and hence any body moving in a circle always tends to move away from the centre of the circle which it describes.*[43]

For here we have a decisive improvement over Galileo's idea of inertial motion, which he conceived as motion parallel to the surface of the Earth. Descartes, by contrast, realized that there was no circular inertia, only rectilinear inertia: even when a body is moving in a circle, "each of its parts individually tends always to keep moving along a straight line"—the tangent to the curve—so that "the action of these parts—i.e., the tendency they have to move—is different from their motion."[44] This is a momentous step, for it enables Descartes to give a completely different kind of analysis of motion. Instead of trying to understand the dynamics of the motion of a slingshot by looking at the motion over time, Descartes's idea was to look instead at the balance of the instantaneous tendencies at an instant: since the normal motion of the stone would be in a straight line, the circular motion produces a centrifugal force, and this is balanced by the action of the sling, which can be felt as a tension in the cord. Thus a curved path over time is broken down into a sequence of "freeze frames" in which all motions are rectilinear. This corresponds closely to Descartes's method in his *Geometry*, where complicated curves can be analyzed in terms of the relations of x's and y's, where the x's and y's stand for straight line segments. (It is also an instance of the more general method announced in the *Discourse on the Method* of 1637, of breaking down any problem into simpler elements first, before proceeding to a solution.)

These notions—that instantaneous motion is a state, that motion is in itself rectilinear, and that forces can be reduced to instantaneous actions—constitute what I see to be the decisive foundational contributions that Descartes makes to natural philosophy. They are not to be found in his predecessor, Isaac Beeckman, from whom he derives so much of his analysis of motion, including the law of inertia, as I have argued elsewhere.[45] Particularly interesting with regard to the contrast with Beeckman is Descartes's condescension toward his former mentor's time atomism, and especially with his inability properly to treat the continuity of motion mathematically. Descartes, as I have also argued elsewhere,[46] even if he is indifferent on the question of the composition of the continuum, is committed to the continuity of God's action, as is evident in the justification he gives for his second law, the third rule in *The World*:

> It depends on God's conserving each thing by a continuous action, and consequently on his conserving it not as it may have been some time earlier, but precisely as it is at the very instant that he conserves it. So it is that of all motions, only motion in a straight line is entirely simple and has a nature which may be wholly grasped in an instant. For in order to conceive such motion it suffices to think that a body is in the process of moving in a certain direction, and that this is the case at each determinable instant during the time it is moving. By contrast, in order to conceive circular motion, or any other possible motion, it is necessary to consider at least two of its instants, or rather two of its parts, and the relation between them...Note that I am not saying that rectilinear motion occurs in an instant, but only that everything required to produce it is present in bodies at each instant which might be determined while they are moving...

According to this rule, then, it must be said that God alone is the author of all motions in the world in so far as they exist and in so far as they are rectilinear; but it is the various dispositions of matter which render them irregular and curved. Likewise, the theologians teach us that God is the author of all our actions, in so far as they exist and insofar as they have some goodness, but it is the various dispositions of our wills that can render them evil.[47]

Moreover, Descartes argues in both his *Monde* and *Principles*, that if we understand God to be perfect in such a way that he is not only immutable in himself, but also in such a way that he always acts in an utterly constant and immutable manner, God's being the primary cause of motion also implies a certain conservation law:

> *God is the primary cause of motion; and he always preserves the same quantity of motion in the universe*... Thus God imparted various motions to the parts of matter when he first created them, and he now preserves all this matter in the same way, and by the same process that he originally created it; and it follows from what we have said that this fact alone makes it most reasonable to think that God likewise always preserves the same quantity of motion in matter.[48]

Thus was born the first conservation law of modern physics, one which was rapidly developed into the law of conservation of momentum by Wren, Wallis, Huygens, and Mariotte soon after Descartes's death, a development completed later by Newton, but which also led, more gradually, to the idea of conservation of energy. The capital point that I wish to make here is that the unlikely basis for this fecund idea, as the above passage makes clear, is the idea that God's *conserving matter by a given force is not really distinct from his creating it with the same force at every single instant*: conservation, in short, is nothing but continuous creation.

That idea had already been spelled out by Descartes in his *Meditations* of 1641, when it was published in Latin along with six sets of objections from leading philosophers of the time. The most prolix of these was by Gassendi—almost a line-by-line commentary. Indeed, Gassendi's objections were so voluminous that Descartes instructed his editor to leave them out (advice that was not obeyed), and Gassendi's replies to Descartes's replies grew into another large book. The passage in the Third Meditation and the ensuing exchange between Descartes and Gassendi, is worth repeating:

> I do not escape the force of these arguments by supposing that I have always existed as I do now, as if it followed from this that there was no need to look for any author of my existence. For since every lifetime can be divided into countless parts, each of which in no way depends on the

others, it does not follow from my having existed a short while ago that I must exist now, unless there is some cause which creates me as it were again at this moment—that is, conserves me. For it is quite clear to anyone who attentively considers the nature of time that the same force and action is plainly needed to conserve any thing at each moment it endures as would be needed to create it anew if it did not yet exist. Thus the fact that there is only a conceptual distinction between conservation and creation is another of those things that are made evident by the natural light.[49]

To this Gassendi objected that the parts of a subject's duration, being merely external, make no difference to its creation or conservation. Further, he objected to Descartes that his existence is contingent from one moment to another "not because a cause is required to create you anew, but because there is no guarantee that there is not some cause present within you which might destroy you, or that you do not have some infirmity within you which would finally result in your demise."[50]

This last objection contains the kernel of Gassendi's difference of opinion with Descartes, and it is surely the original of the words that Malebranche puts in the mouth of Aristes: while Gassendi regards it as necessary for every existent to have a cause that brings it into existence, he does not see it as needing any cause to keep it in existence. Once a substance is created, then (to use Gassendi's metaphor) the waters of time flow past it until some further cause takes it out of existence. To this Descartes replies:

When you deny that *we continually need the influence* [literally, "inflow"] *of the first cause for our conservation,* you are denying a thing which all the Metaphysicians affirm as self-evident, but which the uneducated often fail to think of, because they attend only to causes of *coming to be* [*secundum fieri*], but not to those of *being* [*secundum esse*]. Thus an architect is the cause of a house and the father of his child only in the sense of being the causes of their coming into being; and hence, once the work is completed it can remain in existence quite apart from the cause in this sense. But the sun is the cause of the light which it emits, and God is the cause of created things, not only *secundum fieri,* but also *secundum esse,* and so he must always influence [literally, "flow into"] the effect in the same way in order for it to be conserved.

And this is clearly demonstrated by what I explained about the independence of the parts of time, which you try in vain to evade by proposing "the necessity of the sequence which exists among the parts of time" *considered in the abstract.* It is not this that is at issue here, but rather the time or duration of the enduring thing, and you will not deny that the individual moments of this time could be separated from those next to them, that is, that the enduring thing could at any single moment cease to exist.[51]

Here Descartes makes clear the extent to which his doctrine of continuous creation is indebted to Augustine, not only using the latter's example of the architect, but even his terminology. Nevertheless, like Augustine, although he attributes all true creative activity to God, he does not go so far as to deny actions to souls (as the above quotation from *The World* concerning the dispositions of our wills indicates); and although God is the primary cause, Descartes does not deny real secondary causes, such as the collisions of surrounding material bodies that prevent a body from following its God-given inertial motion. Thus, he is no more an occasionalist than Augustine. But by the same token, since Descartes is just as committed as Augustine is to God's *continuous* creative action, and thus to the continuous existence and duration of created things, it cannot be said that his instantaneism commits him to the existence of time atoms any more than Augustine's claim that the world would not last for an eyeblink without God's governance commits him to that view.

CONCLUSION

We have seen that Descartes's bodies, like Augustine's, have a continuous duration for as long as God chooses to continuously create them. We have also seen that for Descartes a body's motion does not consist in its simply being in one place at one instant, and another at a subsequent instant, as it would on a time-atomistic rendering: a body in motion is distinguished by its instantaneous state of motion, with a corresponding force of motion, its action, *conatus*, or tendency to move. But we have also seen that it is God who supplies this action directly, by the force of his sustaining the body in its state of existence. For a Cartesian body consists only in extension, and there is nothing in the nature of a body as an extended thing to support such a force. Therein lies a tension in Cartesian philosophy. For on the one hand, Descartes is a vociferous opponent of substantial forms, and so he would have no commerce with Augustine's seminal reasons in things. In fact, his analysis of the essence of bodies as consisting in pure extension means that he upholds the Platonic thesis of the passivity of bodies in a particularly severe form. On the other, however, he appears not to have denied secondary causes in the form of bodily collisions, nor the ability of mind-body complexes to initiate motion, despite his accentuation of the causal inertness of bodies. But this was taken up by his occasionalist followers who did deny secondary causes. Louis de la Forge (1632–1666) and Claude Clerselier (1632–1666) argued that, given the passivity of matter, the only cause of the communication of motion among bodies consistent with Cartesian principles is God himself. Although these two still allowed a role for finite minds, Geraud de Cordemoy (1626–1684) extended this line of reasoning to all finite substances, arguing that since God is the only source of action in bodies, force must be identified with the action of his will—and created souls, remember, on the Augustinian view, "do not create, but only make use of the forces supplied by God to bring forth what he has already created." No longer, however, can these forces in bodies be identified with substantial forms or seminal principles. Malebranche built on these arguments, and supplemented them with arguments of the nominalists against the intelligibility of a necessary connection between cause and effect. Here we have a full doctrine of occasionalism, with plenty of parallels with that of al-Ghazālī. But Malebranche does not adopt time atomism to arrive at his view. Instead he adapts Augustine to Cartesianism by dispensing with the latter's seminal reasons and substantial forms; and then following his fellow Cartesians in their insistence that since force in bodies is reduced to God's will, there is no other cause but this.

So although Malebranche's thought is constrained by the some of the same epistemic vectors as the Ash'arites', his conclusions are reached as a consequence of the Augustinian-Cartesian doctrine of continuous creation, together with the extreme Cartesian interpretation of the passivity of bodies and the origin of forces in the will of God. From these premises it follows—without having to assume time atomism—that there is no necessary connection between the state of the world at one instant and its states at subsequent instants. But even if, as appears to be the case, *kalâm* is not an immediate source for Cartesianism, the whole European tradition is nonetheless inseparable from the Islamic, just as the Islamic was not isolated from the Indian, so that occasionalism is well characterized by Arun Bala's phrase as originating in a "dialogue of civilizations."

END NOTES

1. Bala and Joseph (2007).
2. See Nadler (2005) for a good account of Cordemoy's occasionalism and time atomism, and for further references.
3. For approachable accounts, see Burgess and Quevedo (2007), and Smolin (2004), respectively.
4. This example was given by Maimonides (1135–1204) in his exposition of *kalam*, and, according to Sorabji (1983, 279) a variant of it had appeared earlier in the writing of the Ash'arites theologian al-Ghazālī (1058–1111). The example of the dye is Maimonides.
5. Malebranche (1997, 448).
6. Sorabji (1983, 297).
7. Fakhry (1958, 36).
8. Ibid., 27.
9. Koran (8: 67 and 46: 24).
10. Fakhry (1958, 30).
11. Sorabji (1983, 297–298).
12. See Pretzl 1931.
13. Ibid.; see also Sorabji 1983, 386 for discussion.
14. MacDonald (1927).

15. According to the *Encyclopedia Britannica*, 2009, the Sautrāntikas also held that there is nonetheless "a transmigrating substratum of consciousness that contains within it seeds of goodness that are in every person."
16. Fakhry (2004, 35).
17. Ibid., 33.
18. Ibid., 34.
19. Ibid., 33.
20. Ibid., 35.
21. Bala (2006, 112–113).
22. Ibid., 386–397.
23. Ibid., 296.
24. Ibid., 371.
25. Ibid., 398.
26. Ibid., 397. The reference is to Isaac Israeli, *de Elementis*, 8a, ll. 11–14; cf. Wolfson (1976, 472–486).
27. Sorabji (1983, 399).
28. Ibid., 396.
29. Ibid.; Aristotle *Physics* 1.5, 188a 22–26; 3.5, 205b 32–33.
30. Sorabji (1983, 371–377).
31. See Wolfson (1969, 234–238), who claims that Autrecourt was well acquainted with the ideas of al-Ghazālī; this issue among others is discussed in the general account of Occasionalism given in Lee (2008).
32. Fakhry (1958, 9).
33. In the *Search for Truth*, Malebranche appeals, in preference to Aristotle, to Augustine, "that great saint [who] recognized that the body cannot act upon the soul, and that nothing can be above the soul, except God" (VI, 2, iii, OC 2: 310); (Malebranche 1997, 446–447).
34. Sorabji (1983, 302).
35. Ibid., 305: he cites *Phaedrus* 245c–246a, *Laws* 896a–897b.
36. The passivity of bodies, in fact, was one of George Berkeley's main premises in denying them any mind-independent existence.
37. *I Corinthians* 3:6–7. See also Sorabji 1983, 303.
38. Augustine, de Gen. ad Lit. 4.12.22; quoted from Sorabji 1983, 303–304.
39. Malebranche (1992, 228).
40. Sorabji (1983, 304).
41. Malebranche (1992, 228).
42. AT VIIIA, 62; Descartes (1988, I, 240).
43. AT VIIIA, 63; Descartes (1988, I, 241).
44. *The World*: Descartes (1988, I, 96).
45. Arthur (2007).
46. Arthur (1988).
47. AT XI, 44–46; transl. Descartes (1988, I, 96–97), with slight revisions.
48. AT VIIIA, 62; Descartes (1988, I, 240).
49. Descartes, *Meditations* 3: AT VII, 48–49; my translation: cf. (Descartes 1988, II, §33).
50. Gassendi, in Descartes, *Objections and Replies*: AT VII, 301; my translation: cf. Descartes (1988, II, 209–210).
51. Descartes, *Objections and Replies*: AT VII, 369–370; my translation: cf. Descartes (1988, II, 254–255) and Arthur (1988, 362).

BIBLIOGRAPHY

Arthur, R. T. W. 1988. "Continuous Creation, Continuous Time: A Refutation of the Alleged Discontinuity of Cartesian Time." *Journal for the History of Philosophy* 26, no. 3, July: 349–375.

———. 2007. "Beeckman, Descartes and the Force of Motion." *Journal for the History of Philosophy*, 45, no. 1, January: 1–28.

Bala, Arun. 2006. *The Dialogue of Civilizations in the Birth of Modern Science.* New York and Basingstoke: Palgrave Macmillan.

———, and George Gheverghese Joseph. 2007. "Indigenous Knowledge and Western Science: the Possibility of Dialogue." *Race and Class* 49: 39–61

Burgess, Cliff, and Fernando Quevedo. 2007. "The Great Cosmic Roller-Coaster Ride." *Scientific American*, November: 52–59.

Descartes, René. 1983. *Oeuvres De Descartes*,11 vols. Edited by Charles Adam and Paul Tannery. Paris: Librairie Philosophique J. Vrin.

———. 1988. *The Philosophical Writings Of Descartes*, 3 vols. Translated by John Cottingham, Robert Stoothoff, and Dugald Murdoch. Cambridge: Cambridge University Press.

Fakhry, Majid. 1958. *Islamic Occasionalism and its Critique by Averröes and Aquinas.* London: George Allen and Unwin.

———. 2004. *A History of Islamic Philosophy.* 3rd ed. (1st ed., 1970). New York/ Chichester: Columbia University Press

al-Ghazālī, Abū Hāmid. 2002. *The Incoherence of the Philosophers (Tahâfut al-falâsifa).* Translated by Povo Marmura. Utah: Brigham Young University Press.

Keown, Damien. 2004. "Sautrāntika." *A Dictionary of Buddhism. Encyclopedia. com,* retrieved from http://www.encyclopedia.com.

Lee, Sukjae. 2008. "Occasionalism." *Stanford Encyclopedia of Philosophy,* retrieved from http://plato.stanford.edu/entries/occasionalism/.

MacDonald, D. B. 1927. "Continuous Re-Creation and Atomic Time in Muslim Scholastic Theology." *Isis* 9: 326–344.

Malebranche, Nicolas. 1992. *Selections.* Edited by Steven Nadler. Indianapolis, IN: Hackett Publishing Company.

Nadler, Steven. 2005. "Cordemoy and Occasionalism." *Journal of the History of Philosophy* 43: 37–54.

Nicolas Malebranche. 1997. *The Search for Truth and Elucidations of the Search for Truth, (De la recherche de la vérité,* 1674–1675). Translated By Thomas Lennon and Paul Olscamp. Cambridge: Cambridge University Press.

Pretzl, Otto. 1931. "Die frühislamische Atomlehre." *Der Islam* 19: 117–130.

Richard Sorabji. 1983. *Time, Creation and the Continuum: Theories in Antiquity and the Early Middle Ages.* Ithaca, NY: Cornell University Press.

Smolin, Lee. 2004. "Atoms of Space and Time." *Scientific American*, January: 66–75.

Wolfson, Harry. A. 1976. *The Philosophy of the Kalam.* Cambridge MA: Harvard University Press, pp. 472–486.

———. 1969. "Nicolaus of Autrecourt and Ghazali's Argument against Causality." *Speculum* 44: 234–238.

5

PRAMANAS, PROOFS, AND THE YUKTI OF CLASSICAL INDIC SCIENCE

Roddam Narasimha

INTRODUCTION

Why, asked Needham in 1971, had modern Galilean science taken birth in Pisa, and not in Patna or Peking, although for some fourteen hundred years previously the two Eastern civilizations had been ahead of Europe?[1] This is a question that Asians today have to answer to themselves, but an equivalent question has been raised several times in India since the early nineteenth century, although not in precisely the same terms. Various explanations have indeed been offered, ranging from political/military (invasions, wars, and political chaos) to sociocultural (Indian philosophical concepts such as *karma* and *maya*, the hierarchical social organization of Indian society into castes), socioeconomic (Marxist ideas about the ownership of the means of production and the class structure of society), and technological (the lack of development of novel weapon systems and associated tactical and strategic thinking). There have also been suggestions that different civilizations may have different agendas; for example, Francis Bacon equated knowledge with power (of man over nature), whereas the *Sāmékhya* philosophers (among others) in India saw knowledge as what led to liberation (of the individual).[2]

The question is complex and there can be no doubt that many of the factors mentioned above are relevant in some sense. However, some of those factors were also operating during those fourteen hundred years before Galileo, when India and China were ahead in science according to Needham. It seems, therefore, that there must be other factors that have not received enough attention. I have elsewhere proposed that one fundamental reason is connected with the fact that different civilizations have defined "reliable knowledge" in different ways, and have developed different ways of acquiring it. In other words, they have followed different epistemologies. The history of science is not linear, as Needham demonstrates; nor, I wish to suggest, is the history of epistemology; so different epistemologies have been more effective at different times. It, therefore, helps to recognize that both histories are checkered.[3] It is possible to argue that the European scientific revolution, which gave birth to the modern science that is now so universally pursued all across the globe, was in part the result of an epistemological revolution. In highly over simplified terms, this revolution could be seen as a fusion of earlier Greek with the then-recent Eastern methods of doing science and technology.[4] I shall return to this point briefly toward the end, but for the present the question that appears to me to be both important and fascinating in this connection is how classical Indic scientists thought, about what went on in their minds; the rest (with apologies to Einstein who said something very similar about God[5]) is details.

Before we proceed further, let us note one striking characteristic of scientific exchanges that have taken place over millennia in the Eurasian supercontinent: physical products and tools (e.g., iron, textiles, silk, gunpowder, and rockets) as well as mathematical or intellectual tools (e.g., epicycles for describing planetary motion, new numeral systems, and methods of calculation) have been frequently traded.[6] Ways of thinking, however, do not change as rapidly, in spite of the checkered history of epistemology mentioned earlier. Astronomy provides strong evidence for this perception. While, for example, the Greeks appear to have borrowed computational techniques and trigonometry from the Babylonians and geometry from the Egyptians, they put these subjects together in their own very interesting ways; the Indians borrowed epicycles from the Greeks but once again used them in novel ways that the Greeks would not have approved of (for example, by allowing the radius of the secondary circles to vary periodically with time). Some tools used by Ptolemy may be borrowed, but the assumptions of Book I of his *Almagest* would be ignored or even rejected. In other words, ways of thinking change less rapidly than the tools we use to assist those *or other* ways of thinking.[7]

I will begin by addressing the question regarding astronomy and then proceed to give a brief description of the concept of *yukti*, which was highly valued by Indian astronomers and mathematicians (but used very widely in India also in a variety of other fields)[8]. In the section "*Yukti* in Ayurveda," we will consider the role it played in Indic medical sciences, such as ayurveda. We shall then attempt to establish connections between these ideas and the concept of instruments of valid knowledge (*pramanas*) that are often the central issue in Indic philosophical systems. In the concluding section, I shall make some comparisons between the notions behind the *pramana* systems in India and classical Greek ideas about axiomatic systems and proofs, and finally make some comments on the relevance of these ideas to the contemporary scientific scene.

ASTRONOMICAL SCIENCES[9]

The most articulate exponent of the epistemology of Indic astronomical science was Nilakantha Somayaji (1444–1545) of the Kerala School. He was not only a distinguished astronomer and mathematician but also an expert in the six classical Hindu schools of philosophy. His objective was to analyse the process by which a validated knowledge system emerges. The process, according to him, was collective in nature, stretching over more than one generation. It chiefly involved the establishment of a "school" or a tradition that trains new disciples in making observations and computations and continually comparing the two. The key components of his strategy were, therefore, scientific observation, computation, comparison, inference, and instruction, culminating in the establishment of a tradition.[10] There were multiple computational schemes or algorithm packages in Indic astronomy called *siddhantas,* which could be rank-ordered in terms of the degree of success achieved by each in comparison with observation. Such rank-ordering among the *siddhantas* could vary over time. When disagreements got too large new *siddhantas* had to be created. So the *siddhantas* were tentative: not invariant in time, not "eternal laws." What mattered was agreement between observation and computation—a philosophy that can be aptly described as "computational positivism."[11] It is now being realized that much misunderstanding of classical Indic astronomical science is due to lack of appreciation of this philosophical approach underpinning it that has motivated the Billard-Pingree controversy.[12]

Although formal deductionist proofs do not play a role, the work of Indic mathematicians, including those of the Kerala school, included numerous ingenious demonstrations—of results varying from the theorem attributed to Pythagoras to the infinite series in the calculus over a circle and a sphere.[13]

THE CONCEPT OF *YUKTI*

The word *yukti* constantly recurs in any classical Indian discussion of the processes of creating or applying knowledge—practical, scientific, or philosophical. Like the word yoga, *yukti* is also derived from the root *yuj,* to yoke (the English verb with which it has an etymological connection)—to unite, put together, associate, and so on.[14] The concept of *yukti* has many diverse, powerful connotations, all connected in some way with the idea of putting things together to reach a conclusion or draw an inference. *Yukti* also stands for skill, ingenuity; it may even be a clever trick. To summarize, *yukti* has come to mean a (chiefly mental) process, device, tool, or expedient, going all the way from common sense at one end to cunning at the other; more generally skillful, ingenious, and smart practice; more fundamentally intelligent, associative, inferential reasoning. When Pingree said that in deriving the infinite series for the sine of an angle (or for pi) Madhava (c. 1400 CE) "relied on a clever combination of geometry, algebra, and a feeling for mathematical possibilities,"[15] and furthermore (this was *not* noticed by Pingree) when Madhava introduced and used the idea of asymptotic expansions, perhaps for the first time ever in the world, we are witnessing an excellent example of the power of *yukti*. At one end of the spectrum, *yukti-kara* was a metallurgical craftsman who mixes alloys, and *yukti-jna* was a perfume-maker who knew what essences to mix and how to create perfumes (according to Monier-Williams). At the other end of the spectrum from craftsmanship, the philosophers also praised the virtues of *yukti*. The word appears first in the *Aitareya-brahmana* (Monier Williams again). Kapila's *Sāmékhya-sūtra* asserted that what is not *yukti* is not worth acceptance, and those who accepted were no different from children and lunatics.[16] There is the famous saying of the Bhagavad-gita, "yoga (which is related to yukti) is skill in action."[17] The Buddhists spoke about skill in means. Madhva dualists maintained that inference is *yukti*. Behind these admiring presentations of *yukti* appears to be a belief or faith in the extraordinary power of ingenuity and associative, inferential reasoning in solving a variety of practical as well as philosophical problems.

In mathematical astronomy the concept of *yukti* stood for skillful computation-based reasoning in a broad sense, and was a highly prized value. Nilakantha praises his scientific hero Aryabhata as a treasure-chest of all *yukti*; Bhaskara II defines (and praises) *bija-ganita* (algebra) as *yukti* in handling unknown variables (*Bija-ganita* 2). After providing a demonstration of the theorem of the diagonal and sides of a rectangle (as Indians called the theorem attributed to Pythagoras), Nilakantha makes the extraordinary statement, "All this is rooted only in *yukti*, not in the *agamas.*" Now an *agama* (what has come down [from the past]) is usually a traditional, sacred text. So Nilakantha is rejecting the notion that "all this" has come to us from some earlier, possibly sacred text, and declaring that it can instead be demonstrated purely by "*yukti*," that is, skillful argument devised by human intelligence. It is possible that he was using the result to illustrate the power of the conclusions that could be drawn by rational argument in the context of a more general controversy raging in India at various times (including his own), regarding the differences between the mythological (*puranic*) and the rational (*siddhantic*) views of astronomy. This possibility appears less far-fetched in the light of the work of Minkowski on cosmological controversies in the early modern times of Nilakantha.[18]

But *yukti* was also a central concept in ayurveda, as we shall see in the following section.

YUKTI IN AYURVEDA

The science of ayurveda is more ancient than that of classical Indian astronomy. Although the text of the foundational Indian treatise in the field, *Caraka Samhita* (often abbreviated to CS henceforth), was probably composed in the first few centuries CE, many of the ideas that are central to the subject date back to at least a few centuries BCE. Caraka himself attributes the content of his text to Atreya, and acknowledges the account left by Atreya's star pupil Agnivesa as the original source of the text he redacted. We know that there was a strong tradition of medical science already in the times of Buddha: his physician Jivaka had clearly become a legendary figure in his own time. As Chattopadhyaya has pointed out, ayurveda may have been the first significant knowledge system built up in India on a physicalist and rational framework.[19] (We are making here an implicit distinction between science and mathematics: in the latter the *Sulba-sutras* [about seventh to eighth centuries BCE], an exposition of ritual geometry, perhaps represents the first extant work.) An excellent account of ayurveda, especially informed from the point of view of modern medicine, has been presented by Valiathan.[20]

Now the *Caraka Samhita* not only describes diseases and cures, but also provides insight into Indic scientific thinking in relation to some of the orthodox Hindu systems of epistemology, and is, therefore, of great value in understanding the history of these ideas in classical India.

The basis of ayurveda is set down in striking terms in the first book of the *Caraka Samhita*. To begin with, it is noted that there are three kinds of medicine, the first relying on the godly (e.g., rituals, prayer, pilgrimage, etc.), the second on *yukti* (diet, drugs, medical practices, treatments such as fomentation, etc.), and the third on the conquest of the mind (such as control of mental processes, usually associated with yogic practices).[21] The system that Caraka sets out to describe is the one that relies on *yukti*. He does not reject the other systems, but makes his preferences clear about what his own approach to the subject is. (It is a matter of pure convenience to refer to Caraka as the author of CS, but it is by no means certain that CS is the work of a single author of that name.)[22] Indeed, for Caraka therapeutic success is founded on *yukti*.[23] He recognizes that there are many schools of medical treatment, and that they will not agree on many aspects of their subject. But all of them do share some fundamental tenets. These are that there are diseases, that they have causes and so can be diagnosed, and there are feasible cures and treatments.[24] Mere knowledge of the ingredients or the materials that go into making drugs is not enough. If the causes are understood, the possibility of treating them by physicalist methods *exists*: that is the strong message.[25]

Dosage, timing are important parts of *yukti*; and *yukti* is needed to determine the purpose of the treatment, its action, the source of its power, and where, when, how, and with what success the drugs perform.[26] It is for this reason that the man of *yukti* stands taller than one who merely knows the stuff of drugs.

Indeed, Caraka's commitment to *yukti*-medicine—or to scientific medicine—goes so far as to say that all human happiness is rooted in good science.[27] He makes a clear distinction between what is made by *yukti* (acquired, gained) on the one hand, and what is associated with birth (congenital?) or the result of the passage of time (age, season) on the other.[28] According to Caraka, all goals of human life excepting liberation, but including virtue, wealth, and desire, can be achieved by *yukti*.[29] *Yukti* is, therefore, something that is the outcome of human effort, based on the view that there are methods by which well-being can be achieved going beyond those that one is born with or experiences with season and age.

Public discussion and debate (*tarka*) is also highly valued in the CS, which again goes so far as to say that any success achieved (say in the treatment of a disease) without the benefit of discussion and debate must be considered just fortuitous.[30]

Incidentally, CS allots a primary role to food: it is responsible for the very life-breath (*prana*), as well as complexion, pleasant voice, intelligence, happiness, sacred as well as secular deeds, and so on: "all of that is founded in food."[31]

I must hasten to add that this does not mean that Caraka has no regard for other methods; for example, the role of the mind in successful treatment was keenly appreciated by him. But the subject that his treatise deals with has to do with methods of diagnosis, treatment of various diseases, use of diet and drugs (including products of vegetable or animal origin), various practices and procedures that are useful in maintaining health or curing disease, and so forth: it was basically physicalist.

In other discussions, Caraka connects *yukti* to the instruments of valid knowledge (*pramanas*) of classical Indic philosophy. At one place, he declares that special knowledge of diseases can come in three ways: instruction from the trustworthy (tradition), eyewitness observation or perception, and inference.[32] Incidentally, it is of great interest—and we shall return to this point later—that these three are precisely the same instruments of knowledge (*pramanas*) accepted by the *sāmékhya* system of philosophy.[33] *Yukti* is essential because it is the intellectual force that yields knowledge of matters arising from a combination of multiple causes, and covers past, present, and future.[34] Observation tells us about the present but is not sufficient in itself, for it does not tell us how to foresee the future. Similarly the past as recounted by a teacher, for example, might describe the results of work already done, but their relevance to the present and the future can once again be assessed only through the methods of *yukti*. *Yukti*, therefore, enables us to go beyond the system advocated by the purely materialistic Lokayata school, according to which observation is the sole *pramana*, or instrument of valid knowledge.

Elsewhere, however, Caraka actually includes *yukti* on an equal footing with the other three instruments—word of the trusted, perception, and inference—mentioned above, as together constituting the set of four *tests* that enable us to separate the real from the unreal.[35] *Yukti*, however, is not accepted as a *pramana* in any of the classical Indic philosophical systems, which see it chiefly as a powerful tool that *helps* inference (*anumana*). Indeed, philosophers of the dvaita tradition said flatly that inference is only *yukti*. (They went on to say that *yukti* yields something that cannot be demonstrated by other ways, and that what is acquired by *yukti* compels acceptance because of the power of argument and demonstration that it carries.[36]) However, even in dvaita philosophy, it is inference that remains a *pramana*, presumably because *yukti* could include matters other than inference.

THE *PRAMANAS*

A central and enduring concern of Indic philosophical systems has in general been the theory of knowledge: they seek to find out what constitutes reliable (more precisely acceptance-worthy) knowledge, and how one may acquire it. The answer, according to them, lies in the instruments of reliable knowledge (*pramanas*). Each philosophical system declares its *pramanas*, and proceeds on that basis to derive its knowledge system.[37]

We have a total of nine major Indic systems (the six Hindu schools, Buddhism, Jainism, and Lokayata) that have made significant contributions to the science of knowledge or epistemology. Of these the Lokayata is an ancient materialistic system that has not survived. It accepted only eyewitness observation (*pratyaksa*) as a *pramana,* and rejected any other source for reliable knowledge. (Incidentally this choice went with a preference for the happy life as a supreme goal, and led to a characterization of the system as hedonist by its critics—but then we know only what its critics say about it, as no original lokayata texts have survived.) It was, however, quickly realized that an exclusive dependence on eyewitness observation would severely limit the scope of our knowledge, in particular making it so presentist that both past and future would have to remain oceans of total ignorance.

Of the six orthodox Hindu systems, the oldest is *sāmékhya*, as it is already mentioned in the Upanishads. It is also what today would be considered rationalist.[38] The school called "nirisvara sāmékhya" declared that there was no *pramana* for God, and was, therefore, nontheist (it carefully avoided taking an atheistic position). The system postulated a duality between nature or insentient primordial matter (*prakrti*), and spirit or the sentient principle (*purusa*). For the *sāmékhyas* the world is real. No matter could be created out of nothing. Nature evolved over time, by internal dynamics rather than by design. Nevertheless, nature could be (unintentionally) kind and helpful: it rains not because nature wants to help mankind, but rainfall is nevertheless most beneficial. *Sāmékhya* shunned making unnecessary hypotheses—an idea that deserves to be called "Kapila's Razor."[39] It is enumerative, and took its name as "numerism,"—for it put "number in the forefront" (*sāmékhyā*= number), as Vijnana-bhiksu explained.[40]

Sāmékhya considers that the goal of knowledge is liberation (*moksa*), and (as we have seen) accepts three *pramanas*: eyewitness observation (*pratyaksa*), inference (*anumana*) and trustworthy testimony (*apta-vacana*).

Among the other systems *nyaya* and *vaisesika* are particularly relevant. The latter, like the Buddhists, accepted only two *pramanas*: eyewitness observation and inference; testimony of any kind was considered part of inference. The *nyaya* system also accepted testimony from the scriptures (*sabda*) and analogy (*upamana*) as *pramanas,* and laid great emphasis on logic and debate. It was probably influenced by ayurvedic principles of knowledge.

What is interesting from this brief account of a vast subject is that all six Hindu orthodox schools, and the Buddhists, adopt perception and inference as *pramanas,* but some add other sources of valid knowledge. None of these seven systems succeeded in completely eliminating the rest: Indic civilization has preferred to live with this pluralism. Also, none of the systems adopted Greek-style axiomatic and deductionist logic as instruments of knowledge; indeed, there is considerable evidence that there was a studied rejection of such methods as sources of truth (we shall return to this subject shortly). Instead, however, all claims to reliable knowledge had to be preceded by a declaration of the precise instruments adopted and used in the generation of the knowledge that demanded acceptance as being reliable.

In the light of this discussion what is striking is that both Indic astronomy and medicine appear closest to *sāmékhya* in their approach to knowledge. Certainly they accept all three *sāmékhya pramanas*. Nilakantha chose to add one more, namely tradition and historical lore (*aitihya*), presumably because past data (e.g., on eclipses) was important in devising effective astronomical algorithms; but this is not materially different from trustworthy testimony (*apta-vacana*). As we have seen, although Caraka includes *yukti* as one of the four *tests* for knowledge at one place in the text, elsewhere he adopts the *sāmékhya pramanas*. To the extent that *sāmékhya* also holds *yukti* in great regard, this difference again cannot be considered material.

There is more direct evidence of the high regard that the *sāmékhya* system was held in by classical Indic scientists. Caraka speaks of a meeting in which Agnivesa, the original author of the treatise, seeks to clear his doubts from Punarvasu (another name for Athreya), "seated in the company of sāmékhya philosophers."[41] The great mathematician and exponent of algebra, Bhaskara II (twelfth century CE), begins his famous text *Bija-ganita* with an invocatory verse that plays on an interesting series of puns: one set of meanings invokes God (*Isvara*), the other number and *sāmékhya* philosophy. Kautilya lists *sāmékhya* as one of the investigative sciences in his *Artha-sastra*.

There is another aspect of *sāmékhya* philosophy that is also relevant. It does not criticize the Vedas but does not consider them transcendental, eternal (*nitya*), superhuman (*apauruseya*), or flawless (*visuddha*) (*Sāmékhya-karika* 2). (For this position and its belief in a real world, among other reasons, the Hindu Advaita Vedanta philosopher, Samkara was highly critical of *sāmékhya* thought.) We have earlier seen that Nilankantha declared that the astronomical-mathematical knowledge that he was expounding was rooted in *yukti*, not in the traditional sacred texts (*agamas*). Caraka similarly said that it is irrelevant or superfluous to bring into discussions of ayurveda unrelated utterances of Brhaspati, Usanas, and other similar sages.[42] (Brhaspati was the guru of the gods, and Usanas of the asuras.) Caraka was also a dualist: everything, he said, is either real or unreal[43]; whereas the Rgveda says, in a famous passage, "There was no real, there was no unreal."[44]

From what we have said above, it is impossible to avoid getting the strong impression that there was a recurring undercurrent of non-Vedic thought in all classical Indic science, and that its roots can be traced to ancient and enduring *sāmékhya*. This undercurrent was *not anti-Vedic*, and many who directly or indirectly praised *sāmékhya* thought did not hesitate to call upon the early sacred texts occasionally (e.g., Nilakantha appealing to *Taittiriya-aranyaka* in *Jyotir-mimamsa*).[45] Perhaps, they considered both necessary for understanding reality in its broadest sense, but this did not diminish their rationalist outlook in scientific matters.

PROOF

What is striking in Indic scientific epistemology is the absence of formal deductive proof, based on a set of axioms considered self-evident in the style made famous by Euclid in Greece. This issue has been widely discussed in the literature on Indic science. It has been pointed out that although "formal" proofs are not part of the Indic tradition, other kinds of demonstration, involving explanations of the rationale behind mathematical results and their justification, have all appeared extensively in the commentaries on the great works. A very recent discussion of these issues appears in the recently published English translation of the Malayalam text *Yukti-bhasa* of Jyesthadeva.[46]

I wish to make three further points here. The first concerns the connections between the style and philosophy of Indic mathematics and the *pramana* theory that we have discussed in the earlier sections. There is no doubt that Indic mathematicians knew of these connections; Nilakantha, for example, was not only a great mathematician but also a master of the six (orthodox) philosophical systems, each of which (as we have seen) had its own set of *pramanas* for acquiring reliable knowledge. The interesting thing is that some of these systems consider explicitly the use of logical methods based on hypotheses, and conclude that the method is unsatisfactory *as an aid in discovering "truth,"* but that it may be useful in identifying inconsistencies. Let me quote two examples:

The first is from Gautama's *Nyaya-sutra* (third century BCE). (We recall that the four *pramanas* of the *nyaya* system are perception, inference, testimony, and analogy.) Here is a brief extract (ii.2.1–6).

> (Comment) [It is said that] the number of pramanas is not four, as [four others, namely] tradition (aitihya), hypothetical reasoning (arth-apatti), probability (sambhava) and non-existence (abhava) also have the character of pramanas.
>
> (Response) [These proposals are] not in contradiction [of the pramanas in nyaya], because tradition is part of testimony (sabda), and hypothetical reasoning, probability and non-existence are included in inference (anumana).
>
> (Comment) [It is said that] hypothetical reasoning cannot be a pramana, because it leads to non-unique conclusions.
>
> (Response) [This happens] because the lack of a unique conclusion is the consequence of [reposing] confidence in unjustified hypotheses.

The hypothetical reasoning mentioned here is similar to derivation from axioms; the point being made is that the conclusions reached depend on the hypothesis (axiom) chosen. This nonuniqueness was clearly a serious objection to the use of axiomatist methods in Indic thinking.

The second part is a continuation of the above argument with respect to a specific example, namely the possible eternality of sound (ii 2.26).

> (Comment) There is doubt about the eternality of sound because of conflicting opinions.
>
> (Response) Sound is not eternal because it has a beginning, is experienced through a sensory organ and is considered artificial...
>
> (Comment) Sound is eternal because there is no evident reason why it should be destroyed.

This is countered by two arguments. Argument 1 proceeds as follows.

> (Response) If lack of evident cause for destruction [for D, say, i.e., for non-eternality] implies non-destruction [non-D, say; i.e., eternality],
>
> Then lack of evident cause for inaudition [for I, say] implies no-inaudition [non-I, i.e., continued hearing of sound].
>
> This is absurd, because there is no constant audition [i.e., non-I is not observed], [as we know] by experience (pratyaksa).
>
> Therefore lack of evident cause for D does not imply non-D.
>
> Therefore the hypothesis is contradicted, and is wrong.

This is similar to the familiar argument that lack of evidence for A should not be mistaken for evidence of non-A.

A second argument is added:

> (Response) In any case the proposition [about eternality of sound] is not true, for sound can be made to cease by touch [in the case of a ringing bell, for example].

The nature of the argument here is revealing. First of all it concerns physics more than mathematics. A hypothesis is made, and it is shown that if that is true, it leads to an absurd conclusion, showing the hypothesis must be wrong. This is like a standard Greek reductio ad absurdum, but with one important difference: the absurdity is demonstrated not so much by logic as by appealing to observational evidence. A second argument is also offered; again observation plays a decisive role, but the observation comes from an *experiment* on a bell or gong: it is constructive, not deductive. The primary role of observation and experience, the first and universal *pramana* among *all* Indic systems, is thus highlighted. This has not been as obvious in other cultures. For example, Plato's position on the role of observation in astronomy has been rather ambiguous.[47] Gilbert Ryle summarized it by saying, "The astronomy and harmonics prescribed [by Plato] for the education of the young guardians are to have no touch with stellar observation or with acoustic experiments."[48] Aryabhata, Nilakantha, Gautama, and Caraka—all of them would have rejected this proposal outright.

A related point concerns *tarka,* whose position in ayurveda has already been discussed. *Tarka* in general stood for debate-logic, characterized variously also as hypothetical, indirect, or supportive reasoning.[49] Matilal quotes Vatsyayana's commentary *Nyaya-bhasya* on the *Nyaya-sutra,* and Uddyotakara on Vatsyayana, defining *tarka* as reasoning based on some a priori principle.[50] Hence such reasoning cannot yield any empirical knowledge, as it is based on assumptions, preconceptions, or unvalidated hypotheses—an argument that would apply also to the axiomatist method, as the axioms themselves cannot be proved. It is presumably for this reason that *tarka* was not considered a *pramana* in any Indian philosophical system.

It is, therefore, clear that the absence of an axiomatist proof in India had to do with the considered objections of Indic philosophers to that way of reasoning. They were wary of the dubious procedure of making a distinction between some propositions (axioms) that were to be considered true without benefit of validation or proof, and others that were deduced therefrom and thus *became* "truths."

We now come to the third major point that I wish to make, but this time from contemporary science. Developments in the late nineteenth and much of the twentieth century have begun to question the privileged position that axiomatist proofs have for long occupied in Western mathematics. First of all there was Goedel's theorem, which showed that the Hilbert program in logical proof-making could not in principle be successfully completed. Second, in spite of much unease among pure mathematicians, the increasingly powerful technology of digital computing has begun to offer "proofs" of propositions that have resisted proof by classical methods (e.g., the four-color theorem). Third, there have been some revolutionary attempts by the followers of the so-called intuitionist school, in particular by the brilliant Dutch mathematician L. E. J. Brouwer, to abandon altogether the Aristotelian principle of the excluded middle as well as the axiomatist method, and resort to constructivist mathematics (although this attempt, in spite of brilliant work on its foundations, has not been spectacularly effective). On the whole, it now begins to look as if there can be more than one "proving culture," as the papers presented at a Royal Society discussion meeting in 2004 eloquently demonstrated.[51]

And then there are computer scientists such as Edward Fredkin and Stephen Wolfram who are saying—not yet to a very receptive audience, it must be admitted—that the universe may be a simple (computer) program. For we know there are extremely simple algorithms that lead to very complex behavior—spanning the range from perfect order to perfect disorder or chaos.[52] And the recent proposal that data-intensive scientific discovery constitutes a "fourth paradigm" for doing science is an echo of computational positivism.[53] Richard Feynman has said that there are two ways of doing physics—the Greek, proceeding from first principles and axioms, and the Babylonian, relating one thing to another (Indic methods, valuing *yukti*, are closer to the Babylonian). Feynman went on to say that he was a Babylonian, because he had no preconceptions about what nature was like or ought to be.

So the first major conclusion from this lengthy argument is that Asian scholars need not be so defensive about not developing formal methods of proof in the East. As I have argued elsewhere, the history of epistemology is checkered, like that of science itself.[54] Who knows what strange combination of deductionist, inferential, and computational methods (and/or others as yet unsuspected) may emerge as a powerful force in the future? The rise of modern science in Pisa and not in Patna or Peking (going back to the Needham question we started this chapter with) was in part the result of what was then a new hybrid system that Francis Bacon proposed in the sixteenth century. (This proposal followed his scathing criticism of Greek methods.) In Bacon's new system, the axioms were no longer self-evident truths or arbitrary assumptions, as they tended to be in the Greek system. Instead they were to be *inferred from experience*. That is exactly what Isaac Newton did in the *Principia*: its first two books stated *inferred* axioms (more familiar as Newton's three laws) and worked out their consequences in strict Euclidean style. But the third book was full of observations, assessments, and inferences supporting the "axioms" of the first two books and their consequences. Newton was clear that he was switching gears in the third book, so he inserted a special section called *Rules of Philosophical Reasoning* before switching; and he kept appealing to these rules in Book 3.

As Hermann Weyl noted, Indians considered number as logically prior to geometry; and he went on to say that Western science had followed that path for nearly three post-Newtonian centuries. Indeed, what happened in the European scientific revolution was not so much the *mathematization* of physics (as is often asserted: that had occurred much earlier in at least one branch of physics, namely, astronomy), but rather (first) its *algebraization*—a process that began in India and West Asia, as Descartes realized when he called algebra "barbarous" (i.e., foreign); and second its numerization, that is, the use of the much more powerful computational techniques that the new Indian numeral system enabled. I believe that it is these epistemological advances, triggered by influences from the East, that were responsible in part for the spectacular progress of science in Europe and elsewhere in the last four centuries.

That spectacular progress dazzled the East—to such an extent that its classical epistemologies virtually collapsed in the nineteenth and twentieth centuries. Whether the reemergence of science in the East, beginning in the twentieth century, will continue in the epistemological footsteps of Baconian science, or will discover new, more powerful syntheses, is not yet clear.

END NOTES

1. In the form stated, this question appears in Needham's Foreword to the proceedings of a conference published in 1971 but held much earlier. See Needham 1971.
2. A detailed comparison of the views of Bacon and the Samkhya school in India appears in Narasimha (2011).

3. Narasimha (2009b).
4. This point is argued at greater length in Narasimha (2009b).
5. In an interview with Esther Salaman, *The Listener,* September 8, 1955; quoted in Clark (1999, 37).
6. Van Kley (2007).
7. For the Egyptian and Babylonian sources of Greek learning see Bernal (1987). Several chapters in this volume consider the role of intercultural dialogue in the development of science. See also Narasimha (2003).
8. For further details see Narasimha (2012).
9. I am summarizing here the more detailed discussion that appears in Narasimha (2009a).
10. For a detailed discussion of these issues in classical Indic science see Narasimha (2007).
11. See Narasimha 2003, 2009b.
12. Kim Plofker has acknowledged this through a personal communication. See also Narasimha 2003.
13. Sarma (2008).
14. The significance of this word, and of other Indian epistemological terms, has also been discussed elsewhere (Narasimha 2007, 2012).
15. Pingree (2003).
16. Sāmékhya-sūtra verse 1.26.
17. Bhagavad-gita verse 2.50.
18. Minkowski (2004).
19. Chattopadhyaya (1991).
20. Valiathan (2003).
21. See CS verse i 11.62. The explanations of each of the three kinds of medicine included in parentheses in the text are given in the immediately following verses.
22. See Chattopadhyaya 1991.
23. See CS verse i 2.15.
24. CS verse iii 8.38.
25. Caraka first defines a *siddhanta* as a conclusion established by causal reasoning after careful observation by experts using different methods. CS verse iii 8.37. There are four kinds of *siddhantas*. Of these the one that is universally accepted (i.e., by all schools) is *sarva-tantra siddhanta*, whose conclusions are that (i) diagnoses exist, (ii) diseases exist, and (iii) feasible, effective remedies exist (See CS verse iii 8.38).
26. CS verse i 26.31.
27. See CS verse i 30.82. The word *vijnana* is today used for science, and has the general meaning of understanding, judgment, skill, and so on, especially in worldly or secular knowledge.
28. CS verse i.11.34.
29. Caraka sees *yukti* as the intellectual force that yields knowledge of matters arising from a combination of multiple causes; that covers past, present, and future; and that helps attain the three goals of human life [namely, *dharma* / virtue, *artha* / wealth and power, and *kama* / desire, passion]. (CS verse i 11.25). Note that *yukti* cannot promise *mukti* / liberation!
30. CS verse viii 2.26.

31. CS verse i 27.346.
32. CS verse iii 4.2.
33. He also links skill in means (*yukti*) and public debate (*tarka*) to inference. See CS verse ii 4.5.
34. CS verse i.11.25.
35. CS verse i 11.17.
36. Dvaita philosophers believe that "the world is real," "accept eye-witness observation," "inference and tradition" as the three *pramanas*, and provide extensive discussions of inference and *yukti*, considering inference as indeed *yukti* (see Pandurangi 2004). They saw *yukti* as providing compelling argument. Samkhya, which is the ancient dualist school, considered that what is not *yukti* cannot be accepted, for to do so would be nothing less than childish or mad (Samkhya-sutra verse 1.26). Caraka was a confirmed dualist: everything had to be either real or unreal (CS i 11.57).
37. See Hiriyanna 1932.
38. References to the original texts can be found in Narasimha 2007, 2009b.
39. He maintains that the postulation of [unnecessary] entities is unacceptable (see Samkhya-sutra verse 5.30).
40. Detailed references for the citations from Sāmékhya texts will be found in Narasimha (2011).
41. CS verse i 13.2.
42. The Caraka Samhita is severely critical of what it calls "adhikam," literally something that is superfluous, hence unnecessary and irrelevant. (CS iii 8.56).
43. CS verse i 11.17.
44. Rgveda x. 129.1.
45. See Narasimha 2012 for discussion.
46. See Sarma 2008; also Narasimha 2012.
47. Plato's ideas on the role of observation have been extensively discussed; their interpretation varies widely among scholars, from those who consider that Plato was only critical of mere star-gazing to those who believe that his view of direct evidence was responsible for the decline of science in Europe. There are also those who think that Platonic worlds are just what modern theoretical physics is about. But the comment in the text remains valid; I do not think any serious Indic scientist or philosopher would have devalued observation.
48. Ryle (1967).
49. Narasimha (2007).
50. Matilal (1999).
51. See Narasimha 2009b.
52. See examples in Wolfram 2002.
53. Hey et al. (2009).
54. Narasimha (2009b).

BIBLIOGRAPHY

Bernal, M. 1987. *Black Athena: Afroasiatic Roots of Classical Civilization*, vol. 1 (The Fabrication of Ancient Greece, 1785–1985). New Brunswick/ London: Rutgers University Press.

Caraka Samhita. 1994. Text with English translation by P. V. Sharma. Delhi: Motilal Banarsidas.

Chattopadhyaya, D. B. 1991. *History of Science and Technology in India.* Kolkata: Firma KLM.

Clark, R. W. 1971/1999. *Einstein: The Life and Times.* New York: Avon Books.

Cooper, J. M. 1997. *Plato: Complete Works.* Indianapolis, IN: Hackett Publishing Co.

Edwards, P. 1967. *The Encyclopedia of Philosophy.* New York: Macmillan.

Hey T., S. Tansley, and K. Tolle. 2009. *The Fourth Paradigm.* Redmond, WA: Microsoft Research.

Hiriyanna, M. 1932. *Outlines of Indian Philosophy.* London: George Allen & Unwin.

Lach, D. F. 1965–1977. *Asia in the Making of Europe,* vols. 1–2. Chicago, IL: University of Chicago Press.

———, and E. J. van Kley. 1993. *Asia in the Making of Europe,* vol. 3. Chicago, IL: University of Chicago Press.

Matilal, B. K. 1998–1999. *Character of Logic in India.* Edited by J. Ganeri and H. Tiwari. New Delhi: Oxford University Press.

Minkowski, C. 2004. "Competing Cosmologies in Early Modern Indian Astronomy." In *Ketuprakasa: Studies in the History of the Exact Sciences in Honor of David Pingree,* edited by C. Burnett, J. Hogendijk, and K. Plofker, pp. 349–385. Leiden: E. J. Brill.

Monier-Williams, M. 1899. *A Sanskrit-English Dictionary.* Oxford: Clarendon Press (reprinted by Delhi: Motilal Banarsidass).

Narasimha, R. 2011. "Culture Views Nature: Bacon and Samkhya Compared." In *Nature and Culture,* edited by R. Narasimha and S. Menon, vol. 14, part 1. History of Indian Science, Philosophy and Culture. Centre for Studies in Civilizations. New Delhi: Munshiram Manoharlal.

———. 2012. "The *yukti* of Classical Indian Astronomy." In *Foundations of Science,* edited by B. V. Sreekantan. Project on History of Indian Science, Philosophy, and Culture. Centre for Studies in Civilizations. Forthcoming.

———. 2009a. "axiomatism and Computational Positivism: Two Mathematical Cultures in the Pursuit of the Exact Sciences." In *Science in India,* edited by J. Narlikar, vol. 13, part 8, pp 89–118. History of Science, Philosophy and Culture. Centre for Studies in Civilizations. New Delhi: Viva Books.

———. 2009b. "The Chequered Histories of Epistemology and Science." In *Different Types of History,* edited by Bharati Ray, vol. 14, part 4, pp. 89–122. History of Indian Science, Philosophy and Culture. Centre for Studies in Civilizations. Delhi: Pearson Longman.

———. 2007. "Epistemology and Language in Indian Astronomy and Mathematics." *Journal of Indian Philosophy* 35: 521–541.

———. 2003. "The Indian Half of Needham's Question: Some Thoughts on Axioms, Models, Algorithms and Computational Positivism." *Interdisciplinary Science Reviews* 28, no.1: 54–66.

Needham, J. 1971. "Foreword." *Science at the Cross Roads.* London: Frank Cass.

Pandurangi, K. T. ed. 2004. *Pramana-paddhatih* (of Jayatirtha). Bangalore: Dvaita Vedanta Studies and Research Foundation.

Pingree, D. 2003. "The Logic of Non-Western Science." *Daedalus*, Fall: 45–53.

Ryle, G. 1967 "Plato." *Edwards 2967*: 314–333.

Sarma, K. V. 2008. *Ganita-yukti-bhasa of Jyesthadeva* (with English translation and supplementary notes). New Delhi: Hindustan Book Agency.

Selin, H., and R. Narasimha. 2007. *Encyclopaedia of Classical Indian Sciences.* Hyderabad: Hyderabad University Press.

Valiathan, M. S. 2003. *The Legacy of Caraka.* Hyderabad: Orient Longman.

van Kley, E. J. 2007. "East and West." In *Encyclopaedia of Classical Indian Sciences*, edited by H. Selin and R. Narasimha, pp. 120–132. Hyderabad: Hyderabad University Press.

Wolfram, S. 2002. *A New Kind of Science.* Champaign, IL: Wolfram Media.

II

SCIENCE, RATIONALITY, AND
INTERCULTURAL DIALOGUE

6

TRADITIONAL KNOWLEDGE AND THE SMALLPOX ERADICATION CAMPAIGN

James Robert Brown

INTRODUCTION

Smallpox was declared eradicated from the world on October 12, 1979. This was slightly less than 13 years after the World Health Organization (WHO) program was initiated. It is unquestionably a wonderful success for humanity as a whole. How did this great achievement come about, given the very different attitudes—cultural, religious, and theoretical—about the disease?

Smallpox eradication presents us with an interesting example to investigate the relation between modern science and traditional knowledge. How was the disease seen by modern science? How was it seen in terms of traditional knowledge? What role, for better or worse, did religion play? What political factors, if any, were involved? Coming to grips with these questions is the main point of this chapter.

TRADITIONAL KNOWLEDGE

A number of terms are used more or less equivalently: *traditional knowledge, local knowledge,* and *indigenous knowledge.* Sometimes they are used more specifically: *traditional environmental knowledge, traditional medical knowledge,* and so on. They are invariably used in contrast with *modern* or *Western science.* Several aspects of the relationship are hotly contended, but most boil down to this question: Does traditional knowledge have anything to contribute to our understanding of nature and to improving the human condition, or should it give way to modern science?

We can give a rough characterization of traditional knowledge along these lines: Traditional knowledge is based on the long-standing social, physical, and spiritual understandings that have informed a society's material survival and psychological or spiritual well-being. The WHO defines traditional medicine as "the sum total of knowledge, skills and practices based on the theories, beliefs and experiences indigenous to different cultures that are used to maintain health, as well as to prevent, diagnose, improve or treat physical and mental illnesses."[1]

These are at best rough characterizations and problems will be immediately evident. For instance, even though "local" and "indigenous" are often used as synonyms, some beliefs are traditional without being of local origin. One can, for instance, find many practitioners of traditional Chinese medicine in downtown Toronto. And some local or indigenous beliefs might be fairly new, not traditional at all. Some theories seem to count as typical Western science, yet were around for a very long time. Ptolemaic astronomy, for instance was, by the time of Copernicus 1500 years old and, thanks to Dante's *Divine Comedy*, intimately linked to spiritual life. It does not seem like an example of traditional knowledge, yet maybe it should count as an instance. The rough characterizations of traditional knowledge that are on offer do not settle the matter.

And just what do we mean by knowledge? The standard characterization of knowledge is that it is justified true belief. Rival theories can't all be true, so the preferred terms should be "traditional belief," "local belief," "scientific belief," and so on. However, rather than argue over the meaning of "true," let us continue to use "knowledge" in a sense that does not imply truth, whatever we mean by it. We can and should set these problems aside. It is best to work with the rough characterization and see how particular examples work out. We can worry about the proper definition of traditional knowledge later. The relevant facts that concern us are that in much of the world, people use traditional knowledge, especially traditional medicine, rather than some form of modern science. Moreover, this traditional knowledge is often widely shared, rather than being in the hands of a small number of experts. Of course, a shaman or midwife will be more skilled than members of the larger society, but the expertise is not as esoteric or formally regulated as is typical of modern science and medicine. Rather than being taught in special educational establishments (universities), traditional knowledge is usually passed on orally from one generation to the next, and it is typically put into practice by ordinary members of the society, not by highly trained experts.

The contrast is evident in the treatment of smallpox. Variolation, a procedure that I will explain below, was practiced by many in a given society. Vaccination, by contrast, required a large and specialized infrastructure to produce the vaccine, to transport it under controlled conditions, and so on. The difference is only one of degree, but it is a significant one.

HOW SHOULD WE VIEW
TRADITIONAL KNOWLEDGE?

When asking how we should view traditional knowledge, one might wonder who are the "we" in the question? Obviously, I must face up to the fact that I am not writing from a completely neutral point of view. Though I will be as neutral as I can, I come form a position that is immersed in and sympathetic to modern science. So the question becomes, how should someone with such a background view traditional knowledge? There are several options. A list that is far from exhaustive might include the following:

Reject Traditional Knowledge as Pseudoscience
Or Superstitious Nonsense

Someone who takes this view is likely to claim that there is no empirical evidence for the claims of traditional knowledge, and that when confronted with a rival theory produced by modern science, the modern account will possess the various theoretical virtues that make a good theory; that is, it explains, predicts, and so on. Anyone who persists in holding traditional beliefs in the face of such evidence is irrational, according to this stern outlook.

Embrace Traditional Knowledge on
aP ar with Modern Science

This view is favored by sociologists of knowledge who deny the very possibility of objective knowledge. It is also accepted by some anthropologists who, think that the first option given above is the only alternative but find it implausible, not to mention, disrespectful.

Ignore Theoretical Claims But Embrace
Useful Technological Results of
Traditional Knowledge

A paradigm example is acupuncture. It is widely used in the West, but only as a technology. The traditional explanation in terms of chi is dismissed and a different account in terms of nerve stimulation causing endorphin release in the brain is accepted. Pharmaceutical companies are constantly on the lookout for traditional remedies based on some local plant, for instance, that they then investigate, refine, and market. The underlying assumption is that a practice might be effective, even if the understanding of why it works is wildly wrong.

By the way, the practice of searching for and utilizing indigenous medicines is one of the principle sources of friction between modern science and traditional knowledge, turning, as it does, on intellectual property rights and bio-piracy.

View Traditional Knowledge as a Rival Theory That Might Be Right. Such a view is about as neutral as possible. It is also a realist view in that it holds that there is at most one true theory; rivals can both be false but cannot both be true (though they might both be partially right).

The list, I'm sure is not exhaustive. None of these approaches seems quite right, though the last is plausible and certainly contains more than a grain of insight. I want to look at the case of smallpox eradication with these different outlooks in mind. To what extent did any of these four outlooks play a role in the eradication program?[2]

The Disease and its Treatment

Smallpox is a virus with two main forms, *variola major* and *variola minor.* The former is much more dangerous, but the latter is a killer, too. Before eradication, the disease killed several million people each year. Those who survived the disease were seriously disfigured and often left blind. Outbreaks of smallpox were throughout the world, but India, which is my main focus, had the most cases, about half of the world total.

Variolation is a process of inoculation, where some pox matter is introduced into the body of a healthy person. Such a person then acquires an immunity to the disease. *Vaccination* is a process that is similar, but instead of using smallpox matter, a similar but milder virus (e.g., cowpox), is introduced into the body. (By the way, viruses were only discovered in the twentieth century.) Immunity to smallpox is acquired. Edward Jenner discovered this process in the late eighteenth century. He called it "vaccination" after the Latin *vacca* for cow.

The different death rates are striking. Smallpox will kill about 30–50 percent of those who contract it. Variolation, when successful, will provide immunity, but can cause death in some cases. However, the death rate is only 2 percent, which is an enormous improvement over no protection. Vaccination only very rarely causes death, about 0.0002 percent of cases. From the statistics alone it would appear that vaccination is by far the best bet. In addition, there is a serious drawback to variolation. The variolated person is contagious for a few days, often leading to serious outbreaks of the disease. Once vaccinations were available, this was the chief reason for outlawing variolation.[3]

THE ERADICATION PROGRAM

Eradication of smallpox was thinkable because the conditions were good. First of all, smallpox is a human disease; there is no animal reservoir, unlike malaria, which lives in mosquitoes. Second, there is an effective vaccine. Third, there were no major cultural impediments, unlike venereal disease or leprosy, which people often disguise. Smallpox looked a much more promising candidate for eradication than earlier failed malaria and yellow fever programs. As it turns out, smallpox is the one and only disease to be eradicated. (However, current programs aimed at polio and guinea-worm disease look promising.)

Edward Jenner had himself proposed eradication in 1801, but nothing came of that. Indeed, it was not until the mid-1950s that serious thought was given to the project. The main impetus came when the USSR proposed eradication at the 11th World Health Assembly, 1958 (and they later donated most of the vaccines in the early years). Finally, the eradication program was initiated by the WHO, January 1, 1967.

The initial plan called for mass vaccination. This does not mean vaccinations for 100 percent of the population. Instead, so-called herd immunity could be achieved at a level of about 80 percent. What this means is that even though some of the population can still catch the disease, it cannot spread fast enough to sustain itself and will eventually become extinct. Unfortunately, this strategy was not a success. In many regions of India, smallpox outbreaks were worse than before the eradication program had begun.

A new strategy was adopted by 1970, known as "surveillance and control." Whenever an outbreak occurs, people are quarantined and all those who might have had contact were vaccinated. This proved to be successful.[4]

MADHAVA-KARA AND SITALA MATA

Sitala Mata is the Hindu goddess of smallpox. She plays a very large role in our story. Before getting to her, something should be said about Madhava-Kara, for the sake of a contrast. Madhava-Kara wrote *Nidana*, an important treatise on disease. It was written in the eighth century and contained extensive knowledge of smallpox, which it classified with chickenpox and measles. These diseases were explained by humors and diet.

The crucial thing about Madhava-Kara's work is that it was a thoroughly biological and naturalistic account. Similar accounts of disease existed in the West, including humoral accounts of smallpox; it was thought to be a "fermentation of the blood," as it was typically described.

Let's turn now to Sitala Mata, who, as I mentioned, is the Hindu goddess of smallpox. Those who accepted her as the cause of smallpox saw the disease as a kind of blessing, or at least a divine intervention. In typical depictions of the goddess, she is riding an ass and has a basket of seeds on her head. The seeds are like smallpox pustules or blisters. In these representations she also has water in one arm and a broom in the other. It is thought that seeds from her basket fall on people, which results in the person getting smallpox. If she washes the seeds, then the person will survive. But if she uses her broom, the person will die. (Needless to say, there are variations on this theme.)

When someone contracts smallpox, appeals are made to the goddess for help. Variolation is performed and ritual prayers are recited during the procedure. The goddess is seen as both the cause of smallpox and its cure. This dual role is crucial. Note the parallel to variolation. The disease of smallpox, through the process of variolation, leads to its own cure, that is to its immunity.

Dates matter. As I mentioned, the naturalistic theory of Madhava is from the eighth century. The Sitala Mata account is from the sixteenth century and became important only in the eighteenth century. This was the time that the Hindu kingdoms had collapsed; they were replaced by Mughals, who in turn were supplanted by the British. Given these dates, which of these two accounts of smallpox should be considered traditional knowledge? If longevity is what matters, then the humoral theory has the stronger claim to be traditional. But if we stress the widespread nature of belief and any role it might play in people's spiritual life, then clearly the Sitala Mata account must be considered traditional knowledge, even though of relatively recent vintage.

A different and perhaps better way to see these issues is in terms of sophisticated versus popular beliefs. This is implicitly what is often at issue in the modern science versus traditional knowledge distinction. It is also—quite mistakenly—often run together with the Europe-Asia distinction. The fact is, the Madhava-Kara account of smallpox in terms of body humors was quite sophisticated, comparable to humoral theories that flourished in medieval Europe. (European accounts typically involved four body fluids: blood, yellow bile, black bile, and lymph. Madhava-Kara had three: wind, bile, and phlegm.) The general idea is that disease results from an imbalance of the humors and that diet can restore the proper balance.

Humoral theories claim that health problems, including smallpox, are caused from within. The Sitala Mata account, like modern germ theories of disease, claim the cause of smallpox is from outside the body. Of course, the kind of external cause is quite different, a supernatural goddess in one case and a germ, a part of the natural order, in the other. Nevertheless, the relative success of the Sitala Mata account is due in part to this.

I confess to using the terms "naturalistic" and "supernatural" rather freely. Of course, this reflects my own background and should be kept in mind. Followers of Sitala Mata may not have made a similar distinction. But we need not dwell. The real key to understanding involves the practice and politics of variolation.

THE POLITICS OF VARIOLATION

The origins of variolation are unknown, but the procedure is widely thought to have started in China. Sitala Mata worship and the practice of variolation were both common in India since the sixteenth century, though, as I mentioned above, it was not until the eighteenth that it became very widely used. Variolation spread through Asia and was adopted through much of Europe. A British aristocrat, Lady Montague, stationed in Constantinople with her diplomat husband in the early eighteenth century, was sufficiently impressed by the practice and she had her family inoculated. (Earlier some of her family had died from smallpox.) She introduced and promoted the practice in England, where it became widespread.

At the very end of the eighteenth century Jenner discovered vaccination, which was an obvious improvement on variolation. In the mid-nineteenth century, the British government outlawed variolation and made vaccination compulsory. They did this in both the UK and in India, which, of course, was a colony at the time. This was not well-received in either. In Britain there were violent reactions, as compulsory vaccination was seen as a heavy-handed measure directed against the poor.[5] In India forced vaccination was also received with violent reactions. It was seen as yet another unjust colonial imposition. It was also perceived as an attack on the Hindu goddess, Sitala Mata. And it did not help the government's case that the smallpox vaccine was derived from cows.

Educated and politically powerful Indians—before and after independence in 1948—favored vaccination. After Independence, variolation was outlawed again and vaccination made compulsory, but still resistance remained. There was a significant difference of opinion between the urban educated and the rural peasants. The latter may have still viewed things in anticolonial terms, but not Nehru and other Indian leaders. They were very keen on modernization, and after independence modernization could be linked to the anti-imperialist USSR, not to the old colonial masters.

One need not be a Marxist to realize that sometimes the ruling classes of two distinct countries may have more in common with one another than they have with the working classes in their respective home countries. And similarly, the working classes in different countries may have more in common with each other than with their own ruling classes.

Usually, the common basis is economic, but not in this case. Nevertheless, much the same was going on in these two countries, when it came to the eradication of smallpox. The British working class and the Indian peasantry both practiced variolation and both were strongly opposed to the vaccination programs that were imposed upon them by law in Britain and in India. And similarly, the ruling classes of the two countries shared a favorable view of vaccination, in spite of not seeing eye to eye on the "benefits" of colonial rule.

In any society, European or Asian, there is an educated or scientific elite, a ruling class, and the general population (and possibly others, e.g., a religious elite). They can be in harmony or hostile to one another. Alliances evolve as conditions change. Currently in the West, there is more or less complete harmony in physics. The governments of Western countries put a great deal of money into physics research, funding the projects that physicists want. The public may not understand what is going on, but they support it. Popular science programs on TV or in newspapers are well-received. Viewers may say that they did not understand it, but they come away with a sense of awe and enthusiasm. The situation is similar in biology, except in the United States where a large segment of the public reject Darwin. By the middle of the twentieth century, the public in the West had come to see vaccines as a stunning benefit for all. Many serious childhood illnesses, including polio, had been rendered impotent.

In nineteenth century Britain, the ruling class and the scientific elite concurred on vaccines. This was sufficient to make them undertake the policy of outlawing variolation and making vaccines mandatory. After that it was a question of coercion, which was not all that successful. Things became vastly easier later when the general public enthusiastically supported the policy. In India, however, there was no general support for such a program at the time of the WHO initiative in 1967. India's scientific and political classes favored vaccination—and with good reason. The general population, however, favored the Sitala Mata account and the practice of variolation—also with good reason.

When I say "with good reason" I mean it—in both cases. Both procedures bring about immunity to smallpox. Variolation is a highly successful method, even though, as already mentioned, it has two drawbacks: People who have been variolated are for a short time carriers of the disease, which they frequently spread. And the death rate among the variolated is 2 percent. However, it is tied to Hinduism, which is important, if one stresses the spiritual aspect of traditional knowledge. Variolation and the worship of Sitala Mata were intimately linked for a great many Indian peasants.

It must be admitted that this is one of the few instances where one could reasonably cite empirical evidence that religious practice and good health are plausibly linked. Of course, religious injunctions have often improved food hygiene leading to better health, but typical claims of specific connections are much less impressive. For instance, it is often claimed that prayer helps the sick to recover faster, but there is not a shred of evidence that this is so. By contrast, those who take variolation and Sitala Mata to be linked can cite plausible evidence in the form of smallpox immunity that their religious beliefs are right. Ultimately, of course, their argument is not good, since variolation works just as well without the ritual prayers to Sitala Mata. Moreover, vaccination is clearly superior, bringing about the same immunity without the same level of deadly side effects. But it would be quite unfair to call holders of the Sitala Mata account superstitious, or to label them pseudoscientists.[6]

STRATEGIES

The initial implementation of the eradication program occasionally resorted to force. A vaccination team would descend on a village and forcefully vaccinate everyone, adults and children alike. This was bitterly resented and was one of the main reasons for the failure to make any headway in the early years. The change in policy from mass vaccination to surveillance and control was crucial in the ultimate success of the program. Vaccination teams only acted on the report of someone infected. An effective system of rewards was set up to ensure that all cases were reported. Then everyone in contact with the infected person would be compulsorily vaccinated. During an epidemic, the population was generally more receptive to the intervention of health workers and more willing to be vaccinated. Moreover, only those who had any contact with infected people were quarantined and compelled to be vaccinated. The sense of a brutal, unwarranted campaign was much reduced.

There was also an important change in icons. In place of Sitala Mata, who was both cause and cure, smallpox was presented as a demon, a wholly evil cause of smallpox. The cure to be found in the bifurcated needle used to vaccinate.

The bifurcated needle was a hugely important innovation. Dipped in the serum, it would hold the right amount to be scratched onto the skin. A person could be trained in its use in only a few minutes and could vaccinate as many as 1500 people per day. The needle was very cheap and used much less vaccine than other methods, thus reducing costs enormously. This simple invention cannot be overpraised.

The switch to surveillance and control and the change in iconography are just two of the many ingredients in the eradication program in India. Of course, there were a great many more. The Indian government, for instance, had to make a choice between putting their limited resources into the smallpox campaign rather than the antimalaria program, which many in the health ministry thought more important.

There is an important fact—as much sociological as it is biological—about the spread of the disease that made the surveillance and containment strategy workable. Smallpox was not distributed evenly through a region, but tended to be located in pockets and moved on to other pockets. For instance, in the year 1967, in one endemic region of India with 2331 towns, only 101 towns were infected throughout the year, and never more than 20 at a time.[7] This meant that the containment strategy was feasible in a way that it would not be if the disease had spread evenly throughout the entire population.

Facts like this should always be kept in mind when thinking about modern science and traditional or local knowledge. Local knowledge is just that—local. Modern science includes a great deal of social knowledge that could play a significant role in tackling medical problems. We tend to focus on the alleged superiority (or lack thereof) of modern science, paying exclusive attention to its medical, biological, or physical theories. This overlooks its occasional reliance on the social sciences and its administrative infrastructure and management skills. These were capacities that were hugely important in the eradication of smallpox.

WHAT CAN WE CONCLUDE?

At the outset, I raised a number of questions involving the relation between modern science and traditional knowledge in relation to the specific case of smallpox eradication. We can now provide some tentative answers.

Why was the eradication program a success? The WHO-led campaign to eradicate smallpox was certainly a success, even though it was implemented in a great variety of cultural contexts where views about the nature of the disease varied widely. How did this come about (at least in India, which had more smallpox cases than any other country)?

The answer, it seems to me, is that it was a victory of reason *and* political power over a recalcitrant but still reasonable population. The Indian government, the civil servants in the Indian Ministry of Health, and the Indian scientific establishment agreed that smallpox eradication was desirable and feasible through vaccination. This is enormously important. We need only think of the attitudes of some governments to AIDS, polio, or tuberculosis, to see that this sort of consensus is not automatic. Moreover, the Indian government had full international support, which meant that the world would not be too critical if the policy were implemented a bit ruthlessly at times.

Is the Sitala Mata account of smallpox and its treatment a pseudoscience? By itself, it would be hard to see any virtues in the Sitala Mata account of smallpox. However, when it is linked to the practice of variolation, it seems to enjoy considerable empirical support. The link would have seemed to many perfectly natural: Sitala Mata is both the cause and the cure of smallpox, and variolation uses smallpox to achieve immunity from it. Anyone inoculated with smallpox matter became immune. The Indian population had enough experience of variolation to know this; it was a well-established fact. It would be wrong, it seems to me, to call this account of smallpox a pseudoscience.

A word about pseudoscience, since it is a problematic notion. One way of characterizing the concept is in terms of various properties that a theory does or does not have. This has the consequence that if a theory is ever a pseudoscience at any time, then it is a pseudoscience at all times. This has the unhappy consequence that Ptolemaic astronomy, Newtonian mechanics, and the Sitala Mata account of smallpox are all pseudosciences, since they are all currently known to be wrong. Yet, each of these, in their day, had considerable evidential support. A much better way to characterize pseudoscience is contextually. A theory is pseudoscientific relative to a community of believers at a given time, if there is available evidence at that time showing that it is false and there is also an evidentially superior theory available at that time. Thus, in the eighteenth century Newtonian mechanics was a perfectly legitimate science. But it would become a pseudoscience in the hands of any group who believe it *today*, given that they have access to the evidence that supports quantum theory and relativity.

The Sitala Mata account of smallpox and its treatment would be a pseudoscience in the hands of any group who believed it when they have access to the current germ theory of disease and all the evidence that supports it. People in rural India from the eighteenth century to the time of the eradication program had access to no such evidence. Whether they had access in the twentieth century is a harder question. Among other things, evidence provided by the hated colonial sources would be reasonably suspect. What they did have was the evidence of variolation practices that were obviously successful. It is important to contrast this with, say, astrological beliefs or the conviction that sex with a virgin will cure AIDS, which have no empirical support whatsoever, and are legitimately called superstitions and pseudosciences. Traditional knowledge, such as the Sitala Mata account of smallpox, may not stand up to a confrontation with modern science, but it is a mistake to think of it as human folly.

Are all theories on a par? Champions of traditional knowledge sometimes fall into thinking that all beliefs are equally true. Or, if "true" is not the right word, then "equally legitimate," or something along those lines. Thus, the germ theory of disease is no better or worse than the Sitala Mata account. Such a view is known as cognitive relativism. Truth in this view is characterized as what is generally accepted by a community.

It is the counterpart of moral relativism, the view that ethical right and wrong are relative to a particular society's norms. *Relativism* is related to social construction and often taken to be a consequence of it. It says that knowledge (scientific, moral) is tied to a group or society ("Polygamy is morally right for them, but wrong for us." "The Big Bang is factually true for us, but for them the world started in a different way, and they're right, too"). For relativists, there is no moral right or wrong, no factual truth or falsehood over and above what is accepted by a particular society. The old saw, "When in Rome, do what the Romans do," doesn't do relativism complete justice, but captures some of its flavor. Truth, relativists claim, is not something transcendent that we all search for—it is instead connected to the here and now, to the practices of particular communities.

If the doctrine of relativism is meant to say that at all times and in all circumstances, every view is as good as every other, then it is certainly rubbish. A great deal has been written about relativism, both cognitive and moral, so there is no need for me to say more.[8] The grain of truth in it is that for any view, there may be some circumstance in which that view is reasonably believed. This, as I argued above, was the case with the Sitala Mata account of smallpox.

What technological transfers were there? The most important transfer of knowledge involving smallpox, we can actually trace. Variolation was invented in Asia, though it is not known precisely when or where. It had been widely used for some time before the practice was transferred to Europe by the early eighteenth century. The Englishman Edward Jenner discovered vaccination at the end of the eighteenth century. His method of vaccination is based on variolation, namely, introducing into the body of a healthy person some material that will induce a mild form of disease. The difference is that variolation uses live smallpox, while vaccination uses live cowpox, but the technique is the same in both. Jenner could not have done his work without full knowledge of variolation. Vaccination, a clearly superior form of inducing immunity, was transferred to Asia in the nineteenth century. There is glory enough for all.

In sum, when investigating the relations between European and Asian science, or between modern science and traditional knowledge (these are not the same thing, though often run together), it is especially useful to focus on particular cases. This is what I have attempted in focusing on the case of smallpox eradication. The success of the eradication program in India was due to many factors: the beliefs of the scientific elite, the beliefs of the political ruling class, the specific way in which the eradication program was carried out, and so on. The interesting contrast was between the germ theory of disease and how immunity in acquired and the rival Sitala Mata account and its use of variolation. The issues were never satisfactorily resolved in the sense of winning people over one way or the other through rational considerations. The success of the small-pox eradication program hung largely on political decisions, propaganda, and at times outright force. Those who held traditional beliefs about smallpox and its treatment were not so much persuaded to adopt a different approach but were cajoled and coerced into it. Other cases of the interaction of modern science and traditional knowledge are sure to be explained in a variety of other ways. The European science–Asian science relation and the modern science–traditional knowledge relation are much too complex to admit of a single pattern. This, I suppose is obvious. The surprises are in the details.

END NOTES

*I am thankful to my research assistant Boaz Miller for being very helpful in finding relevant material, and to Kathleen Okruhlik for conversations that clarified some issues. Arun Bala, Ric Arthur, Leela Joseph, George Joseph, and several other participants at the Singapore conference provided numerous useful comments. Jeff Kochan provided many valuable comments on the penultimate draft. I am grateful to them all.

1. http://www.who.int/mediacentre/factsheets/fs134/en/index.html.
2. There are a great many topics that arise concerning European and Asian science and concerning modern and traditional science. For a discussion of some, see Bala (2006). Apffel-Marglin (1990) specifically discusses these issues in connection with smallpox eradication.
3. Accounts of the nature of the smallpox virus, the history of its spread, and the history of coping with it can be found in Finer (2004), Glynn and Glynn (2004), and Hopkins (2002), and in WHO (1980).
4. The eradication programme is described in WHO (1980), Henderson (1980), and other publications listed in the Bibliography.
5. For a history of the vaccination controversy in Britain, see Williamson (2007).
6. Apffel-Marglin (1990) also dismisses the charge of superstition in her very thoughtful and informative article. However, I think she runs together two things: her distaste for brutal government policy (which she takes to be class war), and her opposition to "logocentric" views, which she identifies with typical Western science. It is an expression of cognitive and moral relativism.
7. WHO (1980), 32.
8. For more on relativism and social constructivism, see Brown (2001).

BIBLIOGRAPHY

Apffel-Marglin, F. 1990. "Smallpox in Two Systems of Knowledge." In *Dominating Knowledge: Development, Culture, and Resistance*, edited by F. Apffel-Marglin and Stephen A. Marglin, pp. 102–144, Oxford: Clarendon Press.

Bala, Arun. 2006. *Dialogue of Civilizations and the Birth of Modern Science.* London: Palgrave Macmillan.

Bhattacharya, Sanjoy. 2007. "Struggling to a Monumental Triumph: Re-Assessing the Final Phases of the Smallpox Eradication Program in India, 1960–1980." *História, Ciências, Saúde Manguinhos* 14, no. 4: 1113–1129.

Brown, James Robert. 2001. *Who Rules in Science: An Opinionated Introduction to the Wars.* Cambridge, MA: Harvard University Press.

Finer, Kim R. 2004. *Smallpox.* Philadelphia, PA: Chelsea House.

Glynn, Ian, and Jennifer Glynn. 2004. *The Life and Death of Smallpox.* London: Profile Books.

Henderson, Donald A. 1980. "Smallpox Eradication." *Public Health Reports* 95, no. 5: 422–426.

Hopkins, Donald R. 2002. *The Greatest Killer: Smallpox in History.* Chicago, IL: University of Chicago Press.

Hopkins, Jack W. 1989. *The Eradication of Smallpox: Organizational Learning and Innovation in International Health.* Boulder, CO, and San Francisco, CA: Westview Press.

Koplow, David A. 2003. *Smallpox: The Fight to Eradicate a Global Scourge.* Berkeley: University of California Press.

Naraindas, Harish. 2003. "Crisis, Charisma and Triage: Extirpating the Pox." *Indian Economic & Social History Review* 40, no. 4: 425–457.

Nicholas, Ralph W. 1981. "The Goddess Sitala and Epidemic Smallpox in Bengal." *Journal of Asian Studies* 41, no. 1: 21–44.

WHO. 1980. *The Global Eradication of Smallpox: Final Report of the Global Commission for the Certification of Smallpox Eradication.* Geneva: World Health Organization.

Williamson, Stanley. 2007. *The Vaccination Controversy: The Rise, Reign and Fall of Compulsory Vaccination for Smallpox.* Liverpool: Liverpool University Press.

Wujastyk, Dominik. 1995. "Medicine in India." In *Oriental Medicine: An Illustrated Guide to the Asian Arts of Healing*, edited by Jan Val Alphen, pp. 19–37. London: Serindia.

———. 2001. "'A Pious Fraud': The Indian Claims for Pre-Jennerian Smallpox Vaccination." In *Studies on Indian Medical History*, edited by Gerrit Jan Meulenbeld and Dominik Wujastyk, pp.121–154. Delhi: Motilal Banarsidass Publishers.

7

THE DIALOGICAL COPERNICAN REVOLUTION: IMPLICATIONS FOR SCIENTIFIC METHOD

Arun Bala

KUHN AND THE COPERNICAN REVOLUTION

Thomas Kuhn begins his seminal work *The Structure of Scientific Revolutions*, first published half a century ago in 1962, with the words, "History, if viewed as a repository for more than anecdote or chronology, could produce a decisive transformation in the image of science by which we are now possessed."[1] Regardless of our valuation of his claim, there is no doubt that Kuhn's study introduced a radically new twist to how relations between the history of science and the philosophy of science came to be conceived. Prior to Kuhn, the dominant tendency among specialists in the discipline had been to contextualize history of science within what they took to be the wider framework of the logic or epistemology of science. The universality of the epistemological was seen as transcending the contingency of the historical. But the influence of Kuhn led many historians and philosophers of science to perceive history as a touchstone against which to test epistemologies of science.

His work has inspired numerous attempts to model the growth of scientific knowledge including exploring the epistemic benefits and drawbacks of the division of scientific labour. It has also produced a wealth of historical case studies that can be used to understand and test accounts of the ways in which scientific progress occurs. However, there has been a lacuna in connecting the methodological concerns that Kuhn addressed with historical studies of the contributions of non-European cultures to modern science. One explanation for the absence of such connections may be that Kuhn himself, despite the inspiration he provided for the multicultural turn in the history of science, did not think that non-European cultures made any significant contributions to science. In his seminal work, the *Structure of Scientific Revolutions*, he writes:

> Every civilization of which we have records has possessed a technology, an art, a religion, a political system, laws and so on. In many cases those facets of civilizations have been as developed as our own. But only the civilizations that descend from Hellenic Greece have possessed more than the most rudimentary science. The bulk of scientific knowledge is a product of Europe in the last four centuries. No other place and time has supported the very special communities from which scientific productivity comes.[2]

To appreciate how Kuhn's concerns about scientific method can be linked to multicultural or dialogical histories, it is important to consider the reorientation to understanding scientific method that he initiated. Central to this reorientation was Kuhn's account of paradigms as specifying not only the theoretical notions of a scientific discipline, but also shaping observational data and defining acceptable norms and practices. Hence, Kuhn maintained that paradigm shifts involved not only mutations of how we conceive nature, but also transformations of our perceptual experiences and our understanding of what constitutes proper scientific methodology. Although Kuhn did not give a precise characterization of what he meant by the term "paradigm"—indeed Kuhn denied that a definition of his central concept was possible[3]—he presented powerful historical arguments, largely based on case studies, to show that not just our conceptions of nature, but our perceptions of nature and scientific method are also paradigm dependent.

Despite the wide impact of Kuhn's views, it is not often appreciated that his notion of paradigms is inspired by an earlier historical study that he made entitled *The Copernican Revolution*. Published five years prior to *The Structure of Scientific Revolutions*, this study looked at the intellectual, religious, and sociocultural factors that conditioned the key episode that led to the birth of modern science. There is evidence that his notion of how scientific revolutions originate, develop, and culminate transposes a pattern that he discovers in the Copernican Revolution to the wider context of scientific revolutions in general.[4] Although he does not introduce the notion of "paradigm" in his earlier work, he clearly recognizes that when historians speak of the "Copernican Revolution" they use the expression in two different senses.

First, they use it to refer to the heliocentric theory in which Copernicus set the Earth to orbit on its axis and revolve around the Sun to deal with problems in the geocentric theory of Ptolemy that took the Earth to be stationary and the Sun to orbit the Earth.

Second, they use the expression "Copernican Revolution" to refer to the process of articulating a whole set of auxiliary assumptions and principles that came to be associated with the Copernican theory to deal with new problems in physics and cosmology raised by Copernicus's astronomical theory. This is the wider Copernican Revolution. Kuhn sees the wider revolution as the accomplishment of a chain of thinkers from Tycho Brahe and Kepler, to Galileo and Newton. Since 141 years separated the publication of Copernicus's *Revolutionibus* in 1543 from Newton's *Principia* in 1684, before the inconsistencies of the Copernican theory with other established theories and observational evidence could become resolved, Kuhn argued that appeal to reason and evidence alone cannot explain the success of the wider Copernican Revolution—it succeeded only because those committed to achieving it were prepared to work with the heliocentric theory over more than a century despite its inconsistency with other established theories, and its conflicts with observed phenomena. It led Kuhn to argue that the final victory of the Copernican theory had to be explained by sociocultural causes rather than intellectual factors.[5]

To appreciate Kuhn's point let us examine the context in which Copernicus first proposed his theory. It was a time when European astronomy was dominated by the Ptolemaic worldview. This assumed that the Earth was at the center of the universe and that the Sun, Moon, and other planets, as well as the stars, revolved around it. The Ptolemaic model was itself framed in association with cosmological, optical, and mathematical views associated with the names of Aristotle, Plato, and Euclid. Aristotle's cosmology gave the physics of motion for the Ptolemaic system, Platonic optics supplied the ontological framework for the naked eye observations that framed the astronomical data supporting the theory, and Euclid's mathematics served as the main computation tool that Ptolemy deployed to predict and explain planetary positions.[6]

The Aristotelian universe was a bifurcated order. Below the Moon—the sublunar region—the universe was in continual change and flux as the elements, earth, air, fire, and water, continually combined and separated as the bodies they constituted coalesced and dispersed. Above the Moon was the superlunar region that extended to the crystalline stellar sphere in which the stars were embedded, and whose movement carried the stars around the Earth. All the celestial objects of the superlunar region were made of a fifth element *quintessence* that gave them the natural propensity to revolve in circular motion around the center of the universe seen as located at the center of the Earth. Moreover, since Aristotle held that a void is impossible, it came to be assumed that air—an element that had a tendency to rise—filled all space up to the fixed stars. However, since Aristotle's notion that the planets moved in circular orbits around the Earth could not be reconciled with observational data, Ptolemy introduced the notion of epicycles and eccentrics in which combinations of various circular motions were used to account for, and predict, planetary positions.

The optical theory associated with Ptolemaic astronomy was dominated by the Platonic view that perception occurred as a result of an emanation from the eye that came to be intercepted by the object. The pattern of the intercepted rays enables the observer to identify the shape, size, and location of the object in the same way as a blind person does with a cane. David Lindberg describes this as an "extramission" theory of vision, since perception occurs by virtue of something issuing from the eye. Indeed Ptolemy himself was to develop the extramission theory into a powerful mathematical model of optics by treating the emanation from the eyes as expanding outward in the form of a cone—the so-called visual cone.[7]

Ptolemy also remained faithful to the idea associated with Euclidean geometry that mathematics was an axiomatic deductive system based upon self-evident first principles. Furthermore, he embraced the Platonic notion that pure mathematical forms could not be embodied in the natural world with absolute precision. It made him satisfied with the notion that combinations of perfect circular motions could be made to fit observational data without being too concerned about whether these models were actually physically realized or realizable.[8] He saw his task simply as "saving the phenomena" – that is, allowing for successful predictions without necessarily providing a "true" model of the universe.

One reason why the Copernican heliocentric theory did not immediately displace the Ptolemaic theory, but had to wait for over 140 years before it came to be generally seen as a viable alternative, is that it came into conflict with the cosmological beliefs that accounted for the physics of celestial bodies. By placing the Earth in heaven where the Sun used to be located, and placing the Sun at the center of the universe where the Earth used to be located, the heliocentric Copernican theory generated problems such as the following. Why do bodies fall to the Earth if it is not at the centre of the universe? Why are the seas not churning, mountains flying, and winds raging if the Earth were rotating at such a great speed that it made one spin every 24 hours? If the Earth is in the heavens, away from the center of the universe, what sense does it make to say that the heavens are unchanging and heavenly objects are made of a substance distinct from objects on Earth?[9]

Another reason why the Copernican heliocentric theory took so long to get generally accepted is that it came into conflict with observations expected on the basis of Platonic optics. First, if the Earth were revolving around the Sun, and traversed vast distances across its orbit, why did the stars exhibit no parallax displacement? Second, if the planet Venus, say, were sometimes close to the Earth and at other times far away on the opposite side of the Sun, as would be expected if they were both revolving around the Sun at different speeds, its size should appear to vary considerably. Yet the size of Venus given to naked eye observation did not change as much as we would expect if the Copernican theory were true.

Finally, even the mathematics deployed to defend the Copernican theory included trigonometric methods, and later the methods of infinite series and the calculus, that went far beyond the techniques deployed by Ptolemy. In fact, Copernicus himself adopted trigonometric approaches, and Newton of course pioneered differential and integrative geometries to solve astronomical problems. All of these computation techniques not only went far beyond Greek mathematics, but their use also generated epistemological concerns that were not resolved until the nineteenth century.

Hence, the Copernican Revolution came to successfully supplant the Ptolemaic framework only after the articulation of new supporting cosmological, optical, and mathematical ideas. However, more than seven score years separate the *Revolutionibus* of Copernicus from the *Principia* of Newton. At the time of its birth, the Copernican theory could be said to have been refuted both from a logical and an empirical point of view—it was born in conflict with both accepted cosmological views and observational results. This goes a long way to explain why Kuhn saw the wider Copernican Revolution from the *Revolutionibus* to the *Principia* as only made possible because external sociocultural factors motivated the defenders of the theory to persist working with it despite its failures until it won out against all obstacles.

LAKATOS AND THE COPERNICAN REVOLUTION

In contrast to Kuhn, and in opposition to his sociocultural explanation for the growth of scientific knowledge, Imre Lakatos developed an alternative account of science that he labeled "the methodology of scientific research programs." Lakatos defines a research program as having four important components—what he terms "its hard core," "its protective belt," and "a set of positive and negative heuristic rules." Briefly, we may characterize these different components of a Lakatosian research program as follows:

1. The hard core is a set of methodological and ontological commitments that define the program;
2. The protective belt is made of an expanding body of auxiliary theories that can be adjusted and modified to bear the brunt of tests that the program has to continually face;

3. The negative heuristic is a set of rules that forbid any modification or alteration of the hard core, so long as we are working within the program; and

4. The positive heuristic is a set of rules that allow us to modify the auxiliary theories deployed in conjunction with the hard core to explain phenomena, so long as these modifications are progressive. By "progressive," Lakatos implies that the modification of the protective belt not only leads to an increase in the scope of the research program, by explaining all the phenomena explained by the research program prior to the modification, but also makes successful novel predictions.

We can illustrate this by considering the Copernican theory. The hard core of the Copernican theory is that the Earth is spinning on its axis and that the planets revolve around the Sun. The Copernican hard core is different from the hard core of the Earth-centered Ptolemic theory that it replaced. For Ptolemy the central assumptions were that the Earth was stationary, and the Sun, Moon, and planets revolved around the Earth.

The protective belt of the Copernican theory involves other auxiliary, less crucial assumptions required to make the theory work. Lakatos discusses the different kinds of modifications that may be made on the protective belt of a research program. First, we may modify or change an auxiliary theory associated with the hard core assumptions in the process of making inferences to explain phenomena. Thus, Kepler rejected Copernicus's claim that planets moved in epicycles and proposed that they moved in ellipses. Did this increase the scope of the research program? Yes, because it could be used to make more precise predictions and, therefore, increased the scope of applicability of Copernican theory. More accurate predictions meant that Kepler's theory fitted the data observed better than Copernicus's original theory. Did Kepler make any novel predictions? Yes, Kepler used his theory to show that each planet's period of revolution around the Sun obeyed two periodic laws (what is now called "Kepler's second and third laws" in contrast to his first law which states that the planets move in elliptical paths around the Sun).[10] Similarly, Newton kept the hard core of Copernicus but added the idea that the planets revolved around the Sun due to the action of gravitational force. He used the gravitation theory to explain Kepler's laws (i.e., widen the scope of the Copernican theory) and also to predict the motion of the tides (novel phenomenon because without the notion of gravity the idea that the Moon influenced the tides would have been deemed absurd).

Another way of modifying the protective belt is to change the theory that is used to interpret sensory stimuli when we make observations—i.e., altering the interpretive theory that delivers "facts." This may occur when a research program faces problems because it seems to be contradicted by "observed facts." Lakatos recognizes that such situations could arise when an inadequate theoretical framework is deployed to interpret sensory stimuli. Concerning such cases he writes:

> The clash is not "between theories and facts" but between two high-level theories: between an *interpretive theory* to provide the facts and an *explanatory theory* to explain them; and the interpretive theory may be on quite as high a level as the explanatory theory...The problem is how to repair an *inconsistency* between the "explanatory theory" under test and the— explicit or hidden—"interpretive" theories.[11]

Such a change of interpretive theory occurred when naked eye observation showed that the apparent size of Venus did not vary as we would expect on the basis of the Copernican theory. Galileo dealt with the problem by adopting a new optical theory in which the perception of a distant object is mediated, not by an emanation from the eye as the Platonists supposed, but by an emanation from the object to the eye. This is the ray theory of light with which we are now familiar. He deployed the ray theory to show that the naked eye cannot reliably judge the size of distant light objects, and that telescopic observations that showed Venus to change in size, as we would expect on the basis of the Copernican theory, are more dependable.

Third, Lakatos also saw mathematical theories as part of the protective belt, and that this belt could be enriched by new mathematical techniques. It leads Lakatos to examine the various stages through which Newton developed his mathematical tools to resolve problems in planetary and projectile motions.[12] He concludes:

> We may appraise research programmes, even after their "elimination," for their *heuristic power*: how many new facts did they produce...We may also appraise them for the stimulus they gave to mathematics. The real difficulties for the theoretical scientist arise rather from the *mathematical difficulties* of the programme than from anomalies. The greatness of the Newtonian programme comes partly from the development—by Newtonians—of classical infinitesimal analysis which was a crucial precondition of its success.[13]

The notion of a research program growing by retaining the hard core, but modifying the protective belt of assumptions to resolve problems, and successfully predicting novel facts in the process, can explain the shift from the narrow Copernican Revolution to the wider Copernican Revolution without appeal to sociocultural factors.[14] Lakatos assumes that the theories of the expanding protective belt—the auxiliary explanatory theories, the observational theories, and the mathematical theories—integral to the defense of the wider Copernican Revolution were mainly created by Brahe, Kepler, Galileo, and Newton.

In short, Lakatos assumes that the theories of the protective belt that culminated in the wider Copernican Revolution ending with Newton were created *de novo* by early modern European thinkers. Over the last 50 years, since the seminal studies of Joseph Needham, there has developed a vast body of literature that renders this assumption questionable. This literature suggests that the new auxiliary theories, interpretive observational theories, and mathematical theories that made possible the wider Copernican Revolution were really articulations and elaborations founded upon theoretical ideas drawn from non-European scientific traditions. We will find that accommodating these discoveries not only requires us to revise our notions of the history of modern science, but also to modify the rational structure of the methodology of scientific research programs offered by Lakatos. This is what we will now proceed to demonstrate.

THE XUAN YE COSMOLOGICAL RESEARCH PROGRAM

Consider the auxiliary theories that Lakatos refers to as constituting the protective belt of the Copernican research program as it developed over time. These include the notion of space as infinite in extent, space as a vacuum, and the heavens as a region of changing phenomena instead of repetitive circular motions. All these new ideas are integral elements of the wider Copernican Revolution. The infinity of space is a necessary component of Newton's law of linear inertia, the vacuum of space is necessary for there to be no resistance to the planets as they orbit the Sun, and the changing heavens include sunspots, comets, and novae that could not be accommodated into the Aristotelian cosmos. But these changes are central features of the new protective belt for the Copernican research program.

There is now evidence that these new assumptions were not generated *de novo* by early modern European thinkers but were incorporated from Chinese cosmological ideas. The transfer of these ideas were possibly mediated by Jesuit missionary-astronomers who arrived in China in the sixteenth century and had, by the end of that century, established themselves in the Astronomical Bureau of the Chinese imperial court. Their elevation into the highest levels of the bureau can be attributed to the better techniques they brought for calendar calculations and predictions of eclipses. Their dominant presence in the Chinese court enabled them to study Chinese astronomical ideas directly and communicate them back to Europe.

One popular theory at the time when the Jesuits entered China was the *Xuan Ye* theory or the Infinite Empty Space theory—the dominant theory accepted by the neo-Confucians. It postulated that the Earth and other heavenly bodies were floating in an infinite empty space. Ge Hong who lived in the fourth century CE attributes the theory to Qi Meng who lived a couple of centuries earlier in the Late Han era. Ge Hong describes the theory as follows:

> [The] heavens were empty and void of substance...The sun, moon and the company of stars float (freely) in the empty space, moving or standing still. All are condensed vapor...It is because they are not rooted (to any basis) or tied together that their movements can vary so much. Among the heavenly bodies the pole star always keeps its place.[15]

The Europeans must have been struck by its radical divergence from the established views of Aristotelian cosmology.[16] An important difference between the Qi Meng and Aristotelian views is that the Chinese theory saw heavenly bodies as subject to change with a history of birth, growth, and decay. In particular, the *Xuan Ye* theory led Chinese astronomers to see the Sun, the Moon, and the stars as formed by condensations from a vaporous substance *qi* that pervaded all space. By contrast, the Aristotelian view saw these bodies as eternal, embedded in rotating crystalline spheres that carried them in perennial circular paths, or paths that could be described as combinations of circular orbits—that is, epicycles. Indeed the Chinese views are closer to current modern views that take stars and planets to have a history, and to have formed out of dispersed matter in empty space, than to the views of the European astronomers who first arrived in China.

Moreover, their view of the heavens as in flux also made the Chinese pay special attention to unusual astronomical phenomena. It led them to document comets, meteors, and meteorites from as far back as 687 BCE. Their records of novae and supernovae can be found on oracle bones as far back as 1300 BCE. They also came to recognize that the Sun was not a perfect sphere, and that its surface was dotted with sunspots that changed and drifted over time. By contrast, European astronomers, even when they saw comets, novae, and supernovae, did not identify them as astronomical phenomena. Under the influence of the Aristotelian view that the heavens were unchanging, they explained these away as atmospheric events of no concern to them. Thus comets were seen in medieval Europe as hot dry exhalations from the Earth that ascended into the heavens, and linked to atmospheric processes such as lightning, thunder, and rainbows.

It is also well-known that observation of sunspots, comets, and novae—a whole class of phenomena unknown to European astronomy prior to the arrival of the Jesuits in China—made their appearance in European astronomy after contact with China. It was also after the arrival of the Jesuits that Europeans came to accept a new cosmology in which space itself became infinite and came to be seen as a vacuum rather than a region filled with air. Moreover, these changes went into the construction of a protective belt for the Copernican research program that made it possible to do away with many obstacles rooted in Aristotelian cosmology.

However, it may be argued that the incorporation of the new phenomena associated with the *Xuan Ye* (Infinite Empty Space) theory could not be deemed a progressive problemshift for the Copernican research program. According to Lakatos a problemshift can be progressive only if it predicts successfully novel phenomena—what he describes as "hitherto unexpected facts." Since the phenomena that supported the Copernican research program against the Aristotelian research program were drawn from the Chinese *Xuan Ye* theory, and its associated body of observational data, they cannot be novel.

In his study *What is this thing called science?*, A. F. Chalmers rejects such a notion of novelty. He maintains that one strong argument in favor of the Copernican theory from the beginning involved its ability to explain many phenomena known since antiquity. These include the retrograde motion of the planets, that planets are brightest when retrogressing, and the fact that Venus and Mercury are always close to the Sun. The qualitative aspects of these observations are easily explained once we assume that the Earth orbits the Sun along with the other planets, and that its orbit lies outside that of Mercury and Venus. By contrast Ptolemy could not explain these observations naturally but only by introducing epicycles designed to fit their observed positions.[17]

Chalmers concludes:

> The Ptolemaic explanation of retrograde motion did not constitute significant support for that program because it was artificially fixed up to fit the observable data by adding epicycles especially designed for the purpose. By contrast, the observable phenomena followed in a natural way from the fundamentals of the Copernican theory without any artificial adjustment. The predictions of a theory or program that count are those that are natural rather than contrived. Perhaps what lies behind the intuition here is the idea that evidence supports a theory if, without the theory, there are unexplained coincidences contained in the evidence.[18]

If we accept Chalmers's argument then it is possible to see the body of phenomena and ideas offered by the Chinese cosmological tradition as also contributing to the protective belt of the Copernican research program in a progressive fashion. They fit in a natural way with the Copernican research program, but do not integrate easily into the Ptolemaic research program. In short, the Copernican research program forms a symbiotic relationship with the Infinite Empty Space Chinese cosmology, in a way the Ptolemaic research program does not. Hence, including the Infinite

Empty Space theory as part of the protective belt of the Copernican research program treats it as an auxiliary theory that is deployed along with the Copernican research program's central assumptions when we explain phenomena.

IBN AL-HAYTHAM'S OPTICAL
RESEARCH PROGRAM

However, as we have seen, Lakatos also envisaged the possibility that a research program may progress by modifying an interpretive theory that delivers observational data. Such an interpretive theory is also a part of the protective belt of the program. The significance of using appropriate interpretive theories to deliver "facts" has been particularly stressed by the philosopher of science Paul Feyerabend. To support his position Feyerabend quotes the following passage from Galileo's *Dialogue Concerning the Chief World Systems*:

> You wonder that there are so few followers of the Pythagorean opinion [that the earth moves] while I am astonished that there have been any up to this day who have embraced and followed it. Nor can I ever sufficiently admire the outstanding acumen of those who have taken hold of this opinion and accepted it as true: they have, through sheer force of intellect done such violence to their own senses as to prefer what reason told them over that which sensible experience plainly showed them to the contrary. For the arguments against the whirling of the earth we have already examined are very plausible, as we have seen: and the fact that the Ptolemaics and the Aristotelians and all their disciples took them to be conclusive is indeed a strong argument of their effectiveness. But the experiences which overtly contradict the annual movement are indeed so much greater in their apparent force that, I repeat, there is no limit to my astonishment when I reflect that Aristarchus and Copernicus were able to make reason so conquer sense that, in defiance of the latter, the former became mistress of their belief.[19]

Feyerabend also adds that Galileo's argument in the passage above is designed to defend the evidence given by the telescope as superior to that of the naked senses. The problem is that the Copernican research program predicts that the apparent sizes of Venus and Mars should change appreciably when they make approaches close to the Earth and reach points far from the Earth. However, naked eye observations do not support such a conclusion, since the changes in sizes observed are much smaller. By contrast, telescopic evidence did support the predictions of the Copernican theory. However, Galileo had no theory of how the telescope worked to support his position. Hence, Feyerabend concludes that Galileo had no rational basis for convincing opponents who rejected the telescopic evidence. It leads Galileo, so claims Feyerabend, to resort to propaganda and trickery to make his case.[20]

However, it is significant that in the passage Galileo is urging us to conquer sense by reason. This is because Galileo actually did have a theory to defend the findings delivered by the telescope. This was the optical theory of the Arabic scientist Ibn al-Haytham—better known in the West as Alhazen. His published work—*Optical Thesaurus*—had a profound influence on European optics immediately after its translation into Latin in the twelfth or early thirteenth centuries. It influenced great medieval European scientists such as Robert Grosseteste and Roger Bacon, as well as leading optical theorists in the Middle Ages such as John Pecham and Witelo. Significantly, Galileo used the Ibn al-Haytham theory to prove that the Moon was not a polished mirror as some Aristotelians maintained.

Ibn al-Haytham developed his mathematical theory of optics as an alternative to the mathematical optics of Euclid and Ptolemy. The latter had assumed the Platonic notion that we see by virtue of an emanation from the eye. Although the emanation theory made a mathematical approach to optical phenomena possible, Ibn al-Haytham argued against its plausibility. In particular, he drew attention to the ability of bright objects to injure the eye. Since an emanation issuing from the eye cannot, by its very nature, damage the eye we must take the injury to be caused by something external. Moreover, to assume that perception results from an emanation issuing from the eye compels us to conclude that material emanated by the eye can fill the whole of space reached by vision—extending even up the sphere of the fixed stars. Ibn al-Haytham argued that this is an incredible claim lacking physical plausibility.

Instead Ibn al-Haytham proposed what has come to be called "the ray theory of light." Ibn al-Haytham's ray theory assumed that the emanation came from the object observed and vision occurred by virtue of its entry into the eye. In particular, he considered the crystalline lens of the eye to be the organ of sensation—not the retina as we currently recognize. This enabled him to develop a mathematical theory of optics that was also physically more plausible.[21]

It was the ray theory that became the basis for an optics that could serve as part of the protective belt and secure telescopic evidence for the Copernican theory against the Ptolemaic theory. Moreover, the telescope also paved the way for more detailed studies of sunspots, comets, and supernovae going far beyond what Chinese science had discovered with naked eye observations. This went even further to subvert Aristotelian cosmology and support the notion of a changing cosmos.[22] Hence, Alhazen's theory also became a part of the protective belt for the Copernican research program.

THE KERALA SCHOOL COMPUTATIONAL
RESEARCH PROGRAM

Finally, let us turn to the new mathematical apparatus of infinitesimals that Lakatos also takes to be a part of the protective belt developed to defend the Copernican research program. It has been argued with some plausibility that the construction of the mathematics of infinitesimals by Newton and Leibniz actually continues a tradition begun by Indian mathematicians in the fourteenth century. There is now some evidence that European mathematicians could have learnt of these discoveries through Jesuit scholars who reached India more than a century before the same discoveries were made in Europe.[23] The Indian mathematicians who made these discoveries belonged to what has come to be labeled the Kerala School of mathematics. The pioneer of the school, Madhava, had anticipated modern discoveries in important areas such as infinite series expansions for trigonometric and circular functions, as well as some rudimentary processes of differentiation and integration in the fourteenth century. These discoveries are now associated with European mathematicians of the seventeenth century culminating with the achievements of Newton and Leibniz.

Even if it were the case that the new developments in European mathematics in the seventeenth century that paralleled the Indian discoveries were made independently, it is still possible to see it as inspired by Indian mathematical discoveries. The infinite series expansions of trigonometric and circular functions, and the infinitesimals of integral calculus, are made possible only because of the computational powers inherent in the Indian place value number system with zero. This number system made it possible for areas to be computed by dividing them into infinitesimal parts and summing them together, and gradients of curved lines to be computed as the limits of a series of infinite terms.[24]

What is intriguing is that the motivations behind the development of these methods by Indian mathematicians were very similar to those that motivated Newton. This is evident when we think of how Indian thinkers of the Kerala School went about computing planetary positions. They involved computing paths of bodies that tended to move in circular orbits, but were made to deviate by forces acting upon them. Their paths had to be calculated as successive approximations of motion under this double constraint.[25]

Hence, the problem that confronted the Indian mathematical astronomers was similar to what confronted Newton more than two centuries later. Newton had to compute the paths of planets moving under two influences—their innate inertial motion that gave them a tendency to move in a straight line with uniform speed, and their attraction to the Sun that caused them to deviate from this path. By contrast, the

Indian mathematical astronomers saw the natural motion of the planets as circular and uniform, but their real motion as also shaped by two attractive centers—what they termed the *mandocca* and *sighrocca* of their orbits—that made them deviate from circular motion. Just as Newton was led to infinite series expansions of trigonometric and circular functions, and also develop the notion of infinitesimals, and concepts of the differential calculus and analysis, the Indian mathematicians were led to similar discoveries—although they never reached the concept of a limit function.

Thus, when Lakatos speaks of research programs as appraised by the stimulus they gave to mathematics, we cannot ignore the contributions of Indian mathematics. When he writes: "The greatness of the Newtonian programme comes partly from the development—by Newtonians—of classical infinitesimal analysis which was a crucial precondition of its success," it should not be forgotten that this became possible only because it built on early developments of Indian mathematics prior to the Kerala School, and may even have been directly inspired by it. Either way the mathematical components of the protective belt that made possible the progressive movement of the Copernican research program were also inspired by Indian mathematical ideas.

FROM SINGULAR RESEARCH PROGRAM TO FAMILY OF RESEARCH PROGRAMS

The influences delineated above show that the wider Copernican Revolution has to be seen as a long drawn-out process that incorporated ideas from many cultures into the protective belt of the Copernican research program. The discussion so far assumed that the process may be seen as the Copernican research program assimilating the Xuan Ye cosmology, Ibn al-Haytham optics, and Indian computational mathematics as auxiliary theories in its protective belt. However, such a description of the historical process is questionable. In the first place these so-called protective belt theories can also be seen as independent research programs. Moreover, they developed as research programs long before Copernicus proposed his theory and in cultures outside Europe. In their own cultural contexts, they had progressed independently of the Copernican hypothesis. Hence, they cannot be simply seen as theories whose sole value is located in their role as shields to protect the hard core of the Copernican research program. Actually, they should be deemed to have their own distinctive hard core principles that cannot be treated as secondary to the hard core principles adumbrated by the heliocentric model of Copernicus.

I would like to suggest that it is more accurate to see the wider Copernican Revolution as the coming together of a family of research programs—the Copernican heliocentric research program, the Xuan Ye cosmological research program, the Ibn al-Haytham optical research program, and the Kerala School computational research program. This family of research programs—we can call it "the heliocentric family of research programs"—displaced the classical geocentric family of research programs: the Ptolemaic astronomical, Aristotelian cosmological, Platonic optical, and Euclidean mathematical research programs.

The notion of a family of research programs to enrich Lakatos's methodological orientation tells us why we need to distinguish, and have distinguished, the narrow and wider Copernican Revolutions. The former is but a single research program, but the wider revolution is the fusion, through a complex process of mutual accommodation, of a number of different research programs. The situation can be described as the bringing into harmony—or consilience—a number of different research programs. Such a process involves modifying a research program not only in the light of inconsistencies with observed data, but also inconsistencies with its associated family of research programs.

The notion of a family of research programs allows us to recognize that in some circumstances science can progress without predicting novel phenomena. We have seen that the Xuan Ye theory allowed the heliocentric model to incorporate phenomena such as sunspots, novae, and comets known to Chinese astronomers. This made the Copernican theory more credible than the Ptolemaic theory, since the former did not assume the bifurcated cosmos of Aristotle. But these phenomena that support the Copernican hypothesis indirectly are not novel phenomena in the sense that they were not known earlier. This makes the Lakatosian view that a research program can progress only if it predicts previously unknown novel phenomena questionable. We have already seen Chalmers make the same point when he referred to the ability of the Copernican theory to naturally explain the retrograde motion of the planets, and the fact that Venus and Mercury never appear far from the Sun, in a fashion the Ptolemaic theory could not. Chalmers also goes on to note that the observation of parallax in the stars was the first confirmation of a really novel prediction of the Copernican theory, but it occurred in the nineteenth century long after the theory had become widely accepted.[26] However, it must be noted that the infinite space theory made it possible to defend the Copernican theory by maintaining that the stars were sufficiently far away for parallax to be undetectable.

Moreover, there could also be novel phenomena that cannot be explained by any one of the research programs in a family but need a combination of research programs. The explanation of perceived differences in size over the course of time of the observations of planets through telescopes illustrates this point.[27] The variations in size can be explained only by using Copernican theory to show that the planets are sometimes close to the Earth and sometimes far away, and using optics to compute their apparent sizes at different distances from the Earth. In effect Copernican theory is conjoined to Ibn al-Haytham's ray theory of light to effect an explanation of observed phenomena. In a sense the observed variations in size add confirmation to both Copernican and Ibn al-Haytham research programs at the same time by showing how each can be made to lend support to the other. This mutual support also suggests that the evidence for one theory also provides evidence for the other indirectly.

Lakatos also treats the hardcore of a research program as inviolate within the program. But this is not faithful to historical judgments most of us would make. For example, most people would be prepared to say that Newton worked within the Copernican research program, since we see the Copernican research program as reaching a triumphant conclusion with Newton. But if we were strict Lakatosians we would say that Newton broke with the Copernican research program when he made the planets orbit the center of gravity of the solar system—a point some distance away from the center of the Sun.

Newton, of course, made this change for mathematical reasons. He had assumed that the planets had an inertial tendency to move with uniform velocity in a straight line, but came to deviate from this natural inclination because of the gravitational pull of the Sun. Since the planets also attract the Sun, the center of their motion cannot be the center of gravity of the Sun. Thus, it is possible to see Newton as making the change because he faced the problem of accommodating his mathematical computational theory that saw both the Sun and the planets as moving under forces of attraction to each other, and Copernican theory that made the Sun the center of the universe. This compromise involved shifting the orbits of the planets to the center of gravity of the solar system.

Thus, modifying Lakatos's methodology of a single research program to accommodate a group of research programs enables us to understand why the Copernican Revolution is often seen in two different ways— a narrow revolution associated with the original theory proposed by Copernicus, and the wider revolution associated with the cosmological, optical, and mathematical ideas that were incorporated to accommodate the Copernican vision. These involved bringing a family of research programs into consilience with each other. Many of the members of the family were also of multicultural provenance. *Contra* Lakatos, such a dialogical approach involving a family of research programs from a plurality of cultures makes it possible to see why the heliocentric revolution of Copernicus triggered a process where research programs grew by drawing in phenomena already integrated within other research programs, by explaining phenomena in conjunction with other research programs, and even by modifying some hardcore principles of their protective belt without betraying the research program.

END NOTES

1. Kuhn (1970, 1). Although Kuhn published his work as an article in the *International Encyclopaedia of Unified Science* in 1962, all references in this chapter will be to its republished ed. of 1970: Thomas Kuhn, *The Structure of Scientific Revolutions* (Chicago, IL: University of Chicago Press, 1970).
2. Kuhn (1970, 167–168).
3. Kuhn's position actually held that paradigms could be identified only by exemplars. See especially chapter 3 in Kuhn (1970).
4. Sharrock and Read (2002) discuss this connection in extensive detail.
5. Chalmers writes:
   The Copernican Revolution did not take place at the drop of a hat or two from the Leaning Tower of Pisa. It is also clear that neither the inductivists nor the falsificationists give an account of science that is compatible with it. New concepts of force and inertia did not come about as a result of careful observation and experiment. Nor did they come about through the falsification of bold conjectures and the continual replacement of one bold conjecture by another. Early formulations of the new theory, involving imperfectly formulated novel conceptions, were persevered with and developed in spite of apparent falsifications. It was only after a new system of physics had been devised, a process that involved the intellectual and practical labour of many scientists over several centuries, that the new theory could be successfully matched with the results of observation and experiment in a detailed way. No account of science can be regarded as anywhere near adequate unless it can accommodate such factors. (1999, 101)
6. Although there were some tensions between the theories, Ptolemy defended his position by essentially arguing that his goal was not to give a real description of the world but to make accurate predictions, what he termed "saving the phenomena." See Goldstein (1997).
7. See Lindberg (1992, 308).
8. The Platonic and instrumentalist orientation of Ptolemy is also evident in the views of Simplicius in the sixth century when he writes in his commentary on Aristotle's *De caelo*:
   Plato lays down the principle that the heavenly bodies' motion is circular, uniform, and constantly regular. Thereupon he sets mathematicians the following problem: what circular motions, uniform and perfectly regular, are to be admitted as hypotheses so that it might be possible to save the appearances presented by the planets. (Quoted in Hetherington 1996, 271)
9. These problems were solved only after the discovery of the concepts of "inertia" and "gravity" deployed so tellingly by Newton.

10. Kepler's three laws may be formulated as follows:
    1. The orbit of every planet is an ellipse with the Sun as one of its foci.
    2. The line joining a planet to the Sun sweeps out equal areas over equal intervals of time.
    3. The square of a planet's orbital period is proportional to the cube of the semimajor axis of its orbit.
11. Lakatos and Musgrave (1970, 129). Emphasis in original.
12. He describes how "Newton despised people who, like Hooke, stumbled on a first naïve model, but did not have the tenacity and ability, to develop it into a research program." Lakatos and Musgrave (1970, 136).
13. Lakatos and Musgrave (1970, 137).
14. Consider a series of theories T1, T2, T3,... within the Copernican research program where each theory results from the addition of auxiliary clauses to a previous theory to accommodate some anomalies or solve some problems. Lakatos would see such a series of theories within a research program as progressive, if each new theory explains more than its predecessors, and is able to successfully make novel predictions. First, there is theoretical progress because each successive theory in the series explains more than the theory it replaces. Second, there is empirical progress because each successor theory in the series predicts novel phenomena corroborated by observation, and therefore leads to the discovery of new features of the world.
15. Ronan (1981, 86–87).
16. In a letter back to Europe in 1595, the Jesuit astronomer Matteo Ricci wrote of a number of "absurdities"—as he called them—that Chinese astronomers believed in. He listed the following absurd beliefs:
    1. The Earth is flat and square, and the sky is a round canopy.
    2. There are not many skies (as the Europeans held by taking each planet to be carried by a solid rotating sphere), but only one sky.
    3. The space between the planets and the stars is not filled with air—it is a void.
    4. There are five elements (earth, water, fire, wood, and metal), and not four (earth, water, fire, and air) as the Europeans held. Moreover, contrary to European views, the Chinese thought the elements could be transformed into each other.
    5. The eclipse of the Sun is caused by the Moon that dims its light as it approaches the Sun.
    6. During the night, the Sun hides under a mountain near the Earth. However, it is striking that some of the things in this list Ricci takes to be absurd given his Aristotelian cosmological views—that there are not ten skies but one, and that the space between the planets and stars is a vacuum not filled with air—have now become a part of modern science. See Needham (1958, 2).
17. Chalmers, (1999, 139).
18. Chalmers (1999, 140–141).
19. Quoted in Chalmer's 151; Feyerabend (1975, 100–101).
20. Feyerabend (1975, 141).
21. For a more extended discussion of al-Haytham's theory see Bala (2006), 85–94.

22. In his study *The Structure of Scientific Revolutions,* Kuhn overlooks al-Haytham's achievement and influence on medieval European optics when he writes:

No period between remote antiquity and the end of the seventeenth century exhibited a single generally accepted view about the nature of light. Instead there were a number of competing schools and sub-schools, most of them espousing one variant or another of Epicurean, Aristotelian or Platonic theory. One group took light to be particles emanating from material bodies; for another it was a modification of the medium that intervened between the body and the eye; still another explained light in terms of an interaction of the medium with an emanation from the eye; and there were other combinations and modifications besides.

23. For a comprehensive discussion of these issues see Joseph (2011), 435–444.
24. See Joseph (2009), 142–155.
25. To appreciate this we have to consider the way the Kerala School of mathematical astronomers approached their problems by adopting a process laid down many centuries earlier by Aryabhata. They saw the nonuniform motion of the planets to be due to their deviation from their natural tendency to move in a circular motion as a result of the attraction of the planets to points in the heavens that they named the *mandocca* and *sighrocca* of the planet. See Sriram (2002), 105 and Girish and Nair (2002), 91. See also Bala (2006), 75–76.
26. Chalmers (1999, 139).
27. Chalmers makes the point:

It is a consequence of Copernicus's suggestion that the earth circulates the sun, in an orbit outside that of Venus and inside that of Mars, that the apparent size of both Venus and Mars should change appreciably during the course of the year. This is because when the earth is around the same side of the sun as one of those planets it is relatively close to it, whereas when it is on the opposite side of the sun to one of them it is relatively distant from it. When the matter is considered quantitatively, as it can be within Copernicus's own version of his theory, the effect is a sizeable one, with a predicted change in apparent diameter by a factor of about eight in the case of Mars and about six in the case of Venus. (Chalmers 1999, 16–17)

BIBLIOGRAPHY

Aryabhateeyam Proceedings. 2002. *Proceedings of the International Seminar and Colloquium on 1500 Years of Aryabhateeyam.* Thiruvananthapuram: Kerala Sastra Sahitya Parishad.

Bala, Arun. 2006. *The Dialogue of Civilizations in the Birth of Modern Science.* New York: Palgrave Macmillan.

Chalmers, A. F. 1999. *What is This Thing Called Science?* Indianapolis, IN, and Cambridge, MA: Hackett Publishing Co.

Feyerabend, Paul K. 1975. *Against Method: Outline of an Anarchistic Theory of Knowledge.* London: Verso.

Girish, T. E., and Radhakrishnan Nair. 2002. "On the Physical Basis of Indian Geometrical Ideas on Planetary Motion." In *Aryabhateeyam Proceedings*, pp. 83–92. Thiruvananthapuram: Kerala Sastra Sahitya Parishad.

Goldstein, Bernard R. "Saving the Phenomena: The Background to Ptolemy's Planetary Theory." *Journal for the History of Astronomy*, 28 (1997): 1–12.

Hetherington, Norriss S. "Plato and Eudoxus: Instrumentalists, Realists, or Prisoners of Themata?" *Studies in History and Philosophy of Science* 27, no. 2 (1996): 271–289.

Joseph, George Gheverghese. 2011. *The Crest of the Peacock*. Princeton, NJ: Princeton University Press.

———, ed. 2009. *Kerala Mathematics: History and Its Possible Transmission to Europe*. Delhi: B. R. Publishing Corporation.

Kanth, Rajani Kannepalli, ed. 2009. *The Challenge of Eurocentrism: Global Perspectives, Policy, and Prospects*. New York: Palgrave Macmillan.

Kuhn, Thomas. 1970. *The Structure of Scientific Revolutions*. Chicago, IL: Chicago University Press.

Lakatos, Imre, and Alan Musgrave, eds. 1970. *Criticism and the Growth of Knowledge*. Cambridge: Cambridge University Press.

Lindberg, David. 1992. *The Beginnings of Western Science: The European Scientific Tradition of Philosophical, Religious, and Institutional Context, 600 BC to AD 1450*. Chicago, IL: University of Chicago Press.

Needham, Joseph. 1958. *Chinese Astronomy and the Jesuit Mission: An Encounter of Cultures*. London: The China Society.

Ravi, Srilatha, Mario Rutten, and Beng-Lan Goh, eds. 2004. *Asia in Europe/Europe in Asia*. Singapore: Institute of Southeast Asian Studies.

Ronan, Colin. 1978, 1981. The *Shorter Science and Civilization in China: An Abridgement of Joseph Needham's Original Text. Vol. 1 and Vol. 2*. Cambridge: Cambridge University Press.

Sharrock, Wes, and Rupert Read. 2002. *Kuhn: Philosopher of Scientific Revolutions*. Malden, MA: Polity Press.

Sriram, M. S. 2002. "Model of Planetary Model in the Kerala School of Astronomy." In *Aryabhateeyam Proceedings*, pp. 105–113. Thiruvananthapuram: Kerala Sastra Sahitya Parishad.

8

HOW INDIGENOUS ARE "INDIGENOUS
SCIENCES?" THE CASE OF
"ISLAMIC SCIENCES"

Ali Paya

HISTORICAL BACKGROUND

In an historical lecture, entitled "On Teaching and Learning," delivered in Calcutta on November 8, 1882, to a group of young Indian Muslim college graduates, Seyyed Jamal al-Din Asad-Abadi (al-Afghani), one of the first and most influential leaders of the Islamic revivalist movement, emphasized, among other things, that

> It is evident that all wealth and riches are the result of science. There are no riches in the world without science, and there is no wealth in the world other than science...If a ruler neglects the dissemination of sciences among its subjects, the harm will revert to that government...The Europeans have now put their hands on every part of the world. The English have reached Afghanistan; and the French have seized Tunisia. In reality this usurpation, aggression, and conquest have not come from the French or the English. Rather it is science that everywhere manifests its greatness and power.[1]

Seyyed Jamal reminded the audience that Islam's positive attitude toward science helped Muslims in the past centuries to gain mastery in many branches of science. However, he lamented the decline of scientific spirit among Muslims in modern times. In particular, Seyyed Jamal was scornful of ulama who were of the view that modern sciences were impure and non-Islamic and, therefore, should not be taught in Muslim communities:

> The strangest thing of all is that our ulama these days have divided science into two parts. One they call Muslim science, and one European science.

Because of this they forbid others to teach some of the useful sciences. They have not understood that science is that noble thing that has no connection with any modernists and traditionalists nation, and is not distinguished by anything but itself. Rather, everything that is known is known by science, and every nation that becomes renowned becomes renowned through science...The Islamic religion is the closest of religions to science and knowledge, and there is no incompatibility between science and knowledge and the foundation of the Islamic faith.[2]

The sentiments expressed by Seyyed Jamal concerning the importance of science for the well-being and greatness of nations on the one hand, and Muslims' rich scientific heritage on the other, have been recurrent themes in the views promulgated by many Muslim thinkers and activists since the early nineteenth century. However, while Seyyed Jamal's recipe for overcoming backwardness in the field of science appears to be astonishingly modern, there have been other powerful voices that have been preaching different approaches.

Apart from some traditional ulama who, as Seyyed Jamal had noted, drove a wedge between what they called "Islamic Sciences" and what they regarded as modern sciences, the following list briefly represents the views of some other influential Muslim writers or scholars who have introduced projects that deal with the problematic of modern science and Islam.

1) An influential idea that is promulgated by scholars such as Seyyed Hossein Nasr and Seyyed Muhammad Naquib al-Attas is the thesis of "sacred science" as an antithesis to modern science. Both writers are against modernist and postmodernist interpretations of science. Both maintain that a proper sacred science can be developed only within a rich realistic metaphysical framework that acknowledges the role of God as the creator and sustainer of the whole realm of being and the source of all values. The views of both writers are informed by doctrines of Muslim mystics such as Ibn Arabi. Both have identified modern science and rationality with positivism. Both advocate development of wisdom (or *Hikma*) in contrast to knowledge or science as defined by positivist research programs.[3]

2) The late Isma'il al-Faruqi's project of "Islamization of knowledge" is another important contribution to the debate on Islam and science in modern time. Some of his specific objectives for Islamization of knowledge were as follows:

a. To create awareness in the *Ummah* (Islamic global community) of the crisis of ideas.
b. To define the critical relationship between the failure of Islamic thought and its methodology.

c. To work for adopting and incorporating comprehensive Islamic methodology in the fields of social sciences and humanities...[and] laying the foundation for the evolution of Islamic social sciences and humanities.

d. To prepare the requisite intellectual cadres to broaden the field of Islamization of knowledge through providing academic supervision, and establishing academic programs for Islamic studies in all fields of contemporary social sciences and humanities.[4]

The ideas of developing a proper methodology for Islamic sciences and also of developing Islamic social and human sciences are among the most popular themes in many of the projects pursued by Muslim writers who advocate the notion of producing Islamic sciences.

3) Some other approaches to developing Islamic sciences, more fundamentalist in their tenor and import, were developed during the presidency of General Zia al-Haq with financial backing from a number of Muslim states, chief amongst them the Saudi Arabia. The apparent aim of these projects was to develop Islamic sciences that were based on a literal interpretation of the Quran and the tradition of the Prophet.[5]

The list of major projects for Islamization of knowledge can be further expanded by including many other entries from a variety of Muslim countries. However, while the literature, in international languages, on these projects in countries such as Malaysia, Egypt, and Pakistan, is rather rich, there is little known about extensive efforts in this area in Iran since the victory of the Islamic revolution in 1979. For this reason, in the next section of this chapter, I shall briefly discuss the views of some of the better-known exponents of the Islamization of science in this country.

ISLAMIZATION OF SCIENCE IN POSTREVOLUTIONARY IRAN

The credit for taking the first serious step toward developing "modern Islamic sciences" in Iran goes to Hojjat al-Islam Monir al-Din Hosseini, a clergyman from Shiraz who had spent many years in Iraq (Najaf Seminary) before returning to Iran in 1970. Hosseini established the Academy of Islamic Sciences in the city of Qom in 1979, with the objective of bringing about a complete paradigm shift with regard to Muslims' understanding of science and technology.

The academy, after the untimely death of its founder in 2000, expanded its activities. Its current head, Mir-Baqeri who is like his predecessor, a cleric (Hojjat al-Islam), maintains that for a science to be regarded as purely Islamic, all its aspects, conceptual, epistemological, and above all

logical and methodological, must be Islamic, that is, it must be informed
and developed according to the teachings of the revelation:

> From an epistemological point of view if one's will is not towards obtain-
> ing God's pleasure then one understands [things] in a different way, one's
> senses perceive [things] differently, one constructs science differently. It is
> an epistemological point that man's will power is present in his senses, his
> theoretical reason, and his intuition, and affects his intuition, [theoretical]
> assessment, and sensitivity. Secondly, the Almighty's *velāyat* (guardian-
> ship) does not accept any restriction. Man's will must operate under the
> guardianship of God.[6]

Another ardent supporter of the project of developing "Islamic Sciences"
is Mehdi Golshani, a well-known physicist and until recently the head of
the influential Institute of Humanities and Cultural Studies in Tehran.
Golshani maintains that modern science is developed within a non-Divine
metaphysics with atheist assumptions. In his view, developing an Islamic
or religious science is possible, provided we put in place an Islamic or religious metaphysics and base our scientific inquiries on Islamic or religious
assumptions.[7]

Among important centers that are promoting the idea of Islamization
of knowledge, the Institute of Islamic Culture and Thought under
directorship of Hojjat al-Islam Ali Akbar Rashad is worth mentioning. The institute, like the Academy and Institute of Humanities and
Cultural Studies, provides research opportunities for young researchers who intend to work on projects concerning Islamic sciences. The
institute also publishes a quarterly, *Qbasāt*, that gives space to views in
defense of the notion of the Islamization of knowledge. The following
excerpt from one of the papers published in a recent issue of the journal,
gives a flavor of the views advocated by the institute. The paper is written by a group of university graduates from different fields as diverse as
physics, medicine, and sociology who represent noncleric proponents
of the idea of Islamization of knowledge. Riyahi in his paper entitled
"Religious Science: Feasibility and Structure" begins his argument by
rejecting earlier definitions of "Islamic Sciences." That is to say, definitions in which such sciences are identified as either traditional Islamic
sciences such as *fiqh* (Islamic jurisprudence), or as sociology and psychology of religion, or as assessment of scientific statements in religious
texts, or as using religious criteria for assessing scientific theories, or as
efforts for importing the set of values and norms approved by religion
into human sciences.[8]

He then goes on to elaborate his own definition:

> The Islamic society in all of its scientific activities—as in all its other activi-
> ties and decision-makings—is committed to the rituals and rules stipulated

by the Holy Law-Maker and maintains that in this way it can achieve its ideal, namely getting closer to God. The outcome of such an activity is a particular science which is different from other sciences in respect of their aims and teloi. The difference is at least as clear-cut as the difference between alchemy and modern chemistry. Scientific activity of such a society cannot be similar to the research activities of those which regard themselves free from any [religious] commitment in the realm of being. First of all, as far as the choice of research topic is concerned, an Islamic society may regard itself responsible to work on particular topics which it considers to be in line with God's blessing. Similarly, it may consider studying and research on other issues as tantamount to getting engaged in futile and unfruitful activities and refrain from them.

The same is true for the research methods. It is quite possible that many of the current research methods, whether for theory construction or data collection or assessment and appraisal of the truth of theories, may not be conducive to the purposes Muslim researchers have in mind or may be in opposition to the value system acceptable to them. In this case, naturally, some part of the activities of the Muslim scientific community will be directed towards finding research methods suitable for its research activities. To all these we should add the effect of the assumptions, beliefs, and inclinations of Muslim scholars which stealthily select, categorise and interpret their findings; approaches and attitudes which are the results of Muslim scholars' prior education and character building in Muslim societies and through an Islamic system of education.

Another influential figure in the Qom Seminary who supports the idea of Islamization of knowledge is Ayatollah Javadi Amoli. In a recent book, the Ayatollah has suggested a seven-point plan for producing empirical sciences. The plan contains suggestions such as to change the notion of "nature" to that of "realm of creation" so as to, presumably, indicate an intelligent object of study; and to emphasize the role of God as the Effective Cause; and to stress the goal of creation that is God's worship and dissemination of justice as the telos of all researches. The other points of the plan are as follows:

> Fourthly, the central pillar of the [scientific] discussion must be narrated evidence such as a Quranic verse or an authentic hadith (traditions of the Prophet or Imams). Fifthly, the discussion should be further confirmed by other narrative evidence from religious sources. Sixthly, the claims that reason or narrative sources on their own are enough must not be made. Seventhly, interpretation of each part of the realm of creation or being should be carried out by considering interpretations of other parts so that the so-called "interpretation of the realm of being by the realm of being," which is akin to the "interpretation of the Quran by the Quran," could be achieved. [This method of interpretation is valid] since each being in the realm of being is a verse, a word and a line in the Divine book of creation.

Having explained his plan, the Ayatollah goes on to emphasize:

> What was said above is based on the premise that reality is absolute, and
> absolute science and knowledge are possible...Therefore the view that
> regards science as a series of non-provable conjectures, or an emphasis on
> relativism and the doctrine that absolute knowledge and science are out
> of man's reach, cannot have any place in this analysis in which we regard
> science to be in step with authentic narrated evidence and arguments from
> authentic religious sources.[9]

While the above representatives of various projects of Islamization of
science emphasize a comprehensive approach concerning all fields of sciences and all types of methodologies and methods, there are other writers who maintain that Islamization of science should be limited to social
and human sciences and their methods and methodologies. Somewhat
like Karl Mannheim who regarded physical sciences and mathematics,
but not the social sciences, to be universally valid,[10] these writers maintain that it is in the realm of human and social sciences that Islamic sciences should be developed.

One of the most active institutions that uphold this approach to
Islamization of science is the Institute of Howzaeh va Daneshgah
(Religious Seminary and University). This institute, as its name indicates,
is interested in forging a close link between religious research centers and
modern universities. On this theme, it has published the largest number
of books in Persian of any Iranian institute that is pursuing the project
of Islamization of Knowledge. It also publishes a scholarly quarterly that
is devoted to developing appropriate methods and methodologies for
Islamic social and human sciences.

The above brief account, though by no means adequate, gives a flavor
of approaches to the thesis of "Islamic science"/ "Islamization of knowledge." In what follows, I shall argue that the position of those who defend
this thesis, in the sense explicated above, is untenable. This is because, as
I shall argue, such position is either based on a category mistake resulting from an inaccurate understanding of how science is developed, or
is informed by a defective philosophy of science that wrongly separates
physical and biological sciences from social and human sciences, or is
not more informative than the well-known, but general realist view that
metaphysical frameworks have an impact upon the range of possibilities
for developing particular scientific theories. My arguments are developed
within a critical rationalist and realist framework.[11]

SCIENCE AND TECHNOLOGY: SIMILARITIES AND DIFFERENCES

Science and technology are both socially constructed entities. However,
despite great degrees of interaction and mutual impact, especially as far

as modern science and technology are concerned, the two are distinct entities. Science, or more generally knowledge, responds to human cognitive needs. Technologies, of all shapes and types, whether computers or books, shoes or ships, houses or banks, democracies or restaurants, serve two main purposes. On the one hand, they respond to human noncognitive needs. Cars, cutleries, chairs, dresses, coins, and many other technologies, belong to this first subcategory. On the other, they act as facilitators for human of cognitive needs. However, they cannot directly respond to human cognitive needs. Telescopes, laptops, glasses, pens, cyclotrons, universities, books, and many other inventions that facilitate human pursuit of knowledge belong to this second subcategory.

Science or knowledge, according to critical rationalists, is objective. It consists of our best conjectures about reality that are expressed in terms of statements. Such conjectures, are supposed to be in the right track toward a true description of reality unless and until they are refuted by either empirical or rational means or both. Knowledge is, and forever remains, conjectural. All knowledge claims, that is, our conjectures about various aspects of reality, whether natural or socially constructed, must constantly be subjected to the most severe tests so that if they do not present a sufficiently true image of reality they get exposed and declared refuted. Truth, therefore, is the aim of knowledge/science. In other words, we strive toward a true picture of reality by constantly trying to eliminate our errors and learn from our mistakes. As for technology, the aim is always pragmatic; that is oriented toward the solution of practical problems.

Knowledge or science claims are general or universal. They are different from both data and information. The latter are about particular entities, processes, events, or contexts. Knowledge/science claims, even if about particulars (e.g., knowledge about the composition of the solar system) are, in principle, generalizable. Moreover, while data and information only provide description, knowledge claims are attempts at explanation.

Knowledge claims should be publicly accessible and assessable. This is the vey meaning of objectivity. Knowledge is, therefore, different from intuitions, flashes of insight, inspiration, and private and personal experiences. Of course, as critical rationalists argue, all these phenomena could and would pave the way to acquiring knowledge.[12] Their role in producing knowledge is vital. In the absence of these capacities, which have a substantial function in creating conjectures, the growth of knowledge would become impossible.

Although scientists are immersed in local cultures and traditions and they all carry their cultural and metaphysical baggage as well as value systems, they do their best, in their bids to understand different aspects of reality, to keep their conjectures free from such external influences to depict reality itself as faithfully as possible. What makes this task humanly possible is the public accessibility and assessability of scientific conjectures. Critical assessment of scientific conjectures, in whatever field of science, whether physical, biological, social, or human sciences, within the limits of human cognitive abilities as well as the knowledge reservoir available to humanity at each point of time, helps conjectures produced by scientists to represent reality itself and not personal or cultural preferences of scientists. Science or knowledge is therefore value-neutral or at least strives to be so.

Now, while to be value-laden is a vice for scientific conjectures that aim to portray reality, whether natural or social, rather than the peculiarities of scientists' upbringing (unless of course this is the very aim of the inquiry in question), for technologies being impregnated with values cherished by their inventors or end-users is not only a virtue but also an indispensable characteristic. Technologies ought to be user-friendly and this means the more they reflect the values and pragmatic preferences of their inventors or users the more acceptable they are.

Scientific conjectures aim to transcend particular contexts and account for the particularities of each context by incorporating initial and boundary conditions in the general body of the theory. It is this extended theoretical entity that, as Pierre Duhem noted, would bear the brunt of the *modus tollens* arrow of potentially refuting evidence.[13] Einstein's general theory of relativity is supposed to be valid throughout the universe, despite the fact that the particular form of the space-time curvature caused by the gravitational field of the black hole in the center of our galaxy is different from the space-time curvature caused by the gravitational field of a quasar. Technologies, however, are context-sensitive; it may not be possible, without proper fine-tuning, to use a technology devised to respond to the needs of people in one particular environment or context, in other contexts. A car made for the cold and wet climate of Europe cannot be used in the hot and dry deserts of Africa without proper fine-tuning and adjustment.

Another notable difference between science and technology pertains to the fact that scientific knowledge is, by and large, cumulative whereas technological know-how is for a large part tacit and noncumulative. Past scientific conjectures that have been successful for a long period and have successfully defeated our best and most effective attempts at falsification, are routinely incorporated as approximations in the succeeding, more explanatory, theories that supersede them. As for technologies, since their know-how is mostly transferred through some sort of master-disciple relationship, in many cases, if the know-how is lost, it would be lost forever, or at least its retrieval would be extremely difficult.[14]

The criteria for judging advances in science and technology are also different. In science the criterion of approaching the ideal of truth about reality provides a rough (and admittedly not yet very well-formalized) measure for progress. In technology and engineering, where, in contrast, the main concern is usually the advancement in making better, more effective instruments—pragmatic considerations gain more prominence.[15]

Technologies, contrary to a view held by a number of writers including Martin Heidegger,[16] do not have essences. They only have functions. Technologies are individuated by means of their functions. Users of technologies could add functions to technologies, or omit them, to make them fit for the purposes they have in mind. For example, a chair could be used as a weapon or an umbrella should a particular user decide to use it in those ways. Of course, the ability of each technology to assume new functions is not limitless. While for science the final arbiter is always reality, for technologies users' tastes and preferences (which together form an important part of users' networks of meaning) are as important as the constraints imposed by reality on the functions each technology provide.

Each specific technology is identifiable as such, only for those who share a network of meaning or collective intentionality in which that particular technology, with its characteristic functions, is recognized. For an indigenous inhabitant of a remote tribe in Amazon's forests, a laptop is only a thing and not a laptop. Slangs only make sense to those who share their meanings.

Earlier it was suggested that the aim of science, in whatever field, is to discover the truth about reality. At the most basic level such truth corresponds to fundamental laws that govern reality at those levels. In the natural sciences what are called "fundamental laws" are our best guesses for capturing the fundamental *laws of nature*. In the social sciences and humanities, we are trying to discover the fundamental laws about human nature. Critical rationalists maintain that human beings share a common core of humanity that distinguishes them from all other entities. This common core is governed by some fundamental laws that, in principle, can be discovered.[17] While the number of fundamental laws that govern nature and the human nature is limited, the majority of laws in all branches of science consist of phenomenological laws. These laws, incidentally, provide a link between theoretical knowledge and technological know-how.

Phenomenological laws are those laws that are used in specific contexts and for particular phenomena.[18] Examples of such laws are Ohm's law, Hooke's law, the laws of fluid dynamics, Fourier's law of heat conduction, classical laws of gases, Coulomb's law, and so on. According to critical rationalists, all phenomenological laws are derivable from fundamental laws either in a direct manner or by means of what is called "approximate derivation." For example, Coulomb's law is a consequence of Maxwell's equations and the Lorentz force for static charges, and the Euler equation for a perfect fluid is a consequence of the fundamental law of dynamics.[19] However, Kepler's law that planets move in ellipses around the Sun can be approximately derived from Newtonian theory.[20]

While fundamental laws introduced by science are idealized and operate under the restriction of *ceteris paribus* clause, phenomenological laws are not universally valid and are subject to initial and boundary conditions of the contexts to which they are applied. These laws, as was suggested above, forge a link between the fundamental laws of science and technological know-how. To better explain this point, I draw on an example discussed by Nancy Cartwright in her book, *How the Laws of Physics Lie.*[21] My take of this example, is of course, radically different from hers, since she does not subscribe to the realist position advocated by critical rationalism, but supports an empiricist approach called "entity-realism." Her approach, as I have discussed elsewhere, would lead to relativism.[22]

As Cartwright explains there are two models for calculating the small signal properties of a typical amplifier, the T-model and the hybrid-π model (See figures 8.1–8.3). The first model substitutes a circuit model for the transistor in the amplifier and analyses the resulting network. The second characterizes the transistor as a set of two-port parameters and calculates the small signal properties of the amplifier.

The important point about the two equivalent models of the amplifier is that they would help scientists and engineers to calculate the values

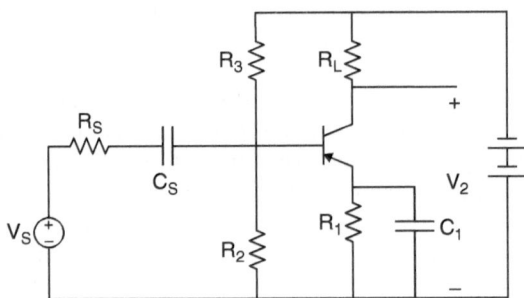

Figure 8.1 Schematic diagram of an amplifier.

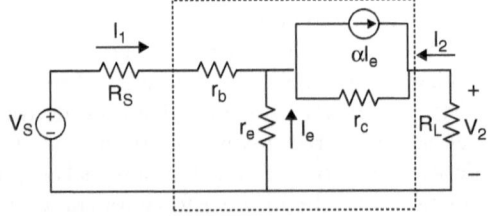

Figure 8.2 T-Model of the amplifier.

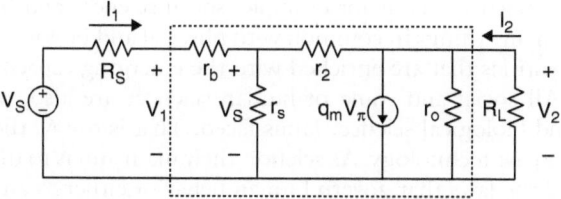

Figure 8.3 Hybrid-π model of the amplifier.

of the transistors' parameters by means of phenomenological laws such as Ohm's law and Kirchhoff's circuit law. These laws are derivable from Maxwell's laws, which represent the fundamental laws in the field of classical electrodynamics. However, the values that are deduced by means of these phenomenological laws are not accurate enough. That is to say, they do not match the real values of the components of a working amplifier. One way to tackle the problem of inaccuracy is to build more sophisticated models that would take into account factors that are omitted from the above two simpler models. Another way to resolve the shortcoming is to make measurements on the actual circuit under study and then use these measured values, and not the theoretically calculated values, for further calculations in the original model. Moreover, in many real cases even after working out various parameters of an entity, whether an amplifier, a car, a gene-splicer, and so on, and even after allowing a good margin of tolerance for the values of these parameters, the optimum performance of the device usually depends on further fiddling with it by those who have accumulated a good deal of know-how and tacit knowledge in relation to the piece in question. Part of this know-how, though never the whole of it, could be used for the purpose of developing better phenomenological laws.

Two more points with respect to the relation between science and technology in general are worth mentioning. The points in question pertain to the interrelations of the four main branches of science, namely, physical, biological, social, and human sciences. According to critical rationalists, contrary to the claim of logical positivists on the one hand, and the advocates of *Geisteswissenschaften* on the other, social and human sciences are neither reducible to physical sciences nor are totally distinct from them. The main branches of science form a continuous spectrum starting from physical sciences and ending with (some branches of) human sciences. Each subsequent branch on this continuum represents a more complex conceptual system in comparison to its predecessor.

This means, among other things, that the subsequent branches have acquired newly emerged capacities that are absent in their predecessors. Or they are dealing with aspects of realities that possess emergent properties not traceable in the aspects of reality investigated by the other earlier branches of sciences. Thus for example, social sciences and humanities, while have many things in common with physical and biological sciences, deal with entities that are enriched with the emerging capacity of intentionality. All social and some of human sciences are also, contrary to physical and biological science, Janus-faced. That is to say, they are part science and part technology. As science, their main aim is to discover and understand the laws that govern human behavior either as an individual or in a community. As for technology, however, their aim is to control and (perhaps) alter such behavior.

Some of the human sciences are also endowed with a capacity that is unique to them and not shared by any other main branches of science. A good part of the specific subject-matter of these fields of humanities is not human collective or individual behavior but the content of other sciences. These second- and higher-order types of knowledge, that is, knowledge about knowledge, combine normative and descriptive-explanatory aspects. This is a feat not accomplished by any other sciences whose subject-matter is first-order phenomena and whose aim is to describe and not prescribe.

A BRIEF CRITICAL ASSESSMENT OF THE CLAIMS MADE BY THE ADVOCATES OF INDIGENOUS (ISLAMIC) SCIENCES

Having explicated the main characteristics of science and technology, we can now address the principal question of the present chapter, namely, the status of indigenous sciences and technologies. Science, as noted above, deals, in the first instance, with universal aspects and features of reality. To achieve this aim, scientists, given their limited cognitive abilities and the restricted range of applicability of their probes and instruments, have no choice but to investigate particular phenomena to discover reality's general and universal secrets. It is in this spirit that economists, for example, strive to find out about the ways human beings behave as *homo economicus*. Having discovered something about this fact, they then combine that general knowledge with particular facts about the particular phenomenon under consideration, for example, the habits and ways of life of the Chinese, to predict, control, or alter their financial/economic activities.

It was suggested earlier that according to critical rationalists, human beings share a common core of humanity. The main elements of this common core consist of a capacity for the creation and construction of meaning; an aptitude to acquire, develop, and critically assess second- and higher-order knowledge; an ability to comprehend moral principles and act according to them; and an adeptness in constructing rules, regulations, and normative measures. Critical rationalists, therefore, reject the idea that human identity is 100 percent socially constructed and human beings enter the world as *tabulae rasae* without any degree of genetic and cultural preprogramed motor and cognitive capacity.

The idea of a common core of humanity and a reality that is independent of human cognitive ability, language and conventions, plus the points about differences between science and technology, provide the ground for dispelling the myth of indigenous sciences.

In what follows, I shall confine my arguments to the claim of the Iranian writers whose views were briefly touched upon in section on "Islamization of Science in Postrevolutionary Iran" above, although similar arguments can be applied to the doctrines of other advocates of the Islamization project in other countries.

One of the theses subscribed to by many of the writers on Islamization of knowledge is rejection of the realist conception of the correspondence theory of truth. Thus for example, Hojjat al-Islam Hosseini the founder of Academy of Islamic Sciences maintains that

> in an Islamic science, properly understood, the notion of correspondence truth is rejected. The correctness of scientific laws [and indeed all human knowledge] is not dependent on their correspondence with reality, but on the sustainability of their usefulness in relation to God's worship. Correctness, in this sense, has a permanent link with truth and value. A value system is a system of aims or goals which are based on either worship of God or of material universe. God's worship and passing through stages of perfection consist of rituals and ways whose clarification is the task of Prophets and saints. The Quran and the Tradition (of the Prophet and Imams) contain the purest knowledge (*ma'ārif*) and in the absence of criteria provided by these sources, recognition of positive or negative applicability of applied sciences is not possible. Therefore, if the applicability of a scientific theory for God's worship is established, its truth will be established.[23]

This alternative criterion suggested by the late Hojjat al-Islam is of course a version of pragmatist theory of truth. But the snag with a pragmatist theory of truth is fourfold: in the first place it is not a proper theory of truth but an epistemological theory, in other words, here, truth is epistemologized; second, it relies for its functioning on the correspondence theory of truth; third, it reduces sciences to technology that is a serious category mistake; and lastly, it leads to a fully fledged relativism in knowledge matters.[24]

Similarly, Ayatollah Javadi's proposal for the Islamization of science does not seem to be conducive to the desired result. He claims, among other things that "the central pillar of the [scientific] discussion must be narrated evidence such as a Quranic verse or an authentic hadith [tradition of the Prophet or Imams]," and that "what lies inside the framework of religious knowledge [i.e., all type of valid knowledge] must be absolute and certain or reliable knowledge." He also thinks that taking scientific theories as conjectures lead to relativism.

But in emphasizing all these points, the Ayatollah, seemingly, forgets that our access to the hidden meanings of the Quran or hadith, and indeed our understanding of any phenomenon, is always limited by our ability to "interpret" texts. In addition, all publicly accessible and assessable interpretations are, and will always be, conjectures.

Moreover to claim, as the Ayatollah does, that all genuine knowledge is Islamic, is not more than stating a truism for those who share the metaphysical underpinning of this claim. After all, for the believers, everything in the realm of being belongs to God and is therefore, in a sense, Islamic or Divine.

Golshani's suggestion for creating an Islamic metaphysical framework as a sure way for producing Islamic science also amounts to nothing more than a pious recommendation with little (if any) practical consequence. The reason is simple. First of all, all metaphysical principles must be interpreted by people and people's interpretations usually differ. Suffice it to note how many different Islamic sects are claiming to be in possession of *the* authentic understanding of Islamic tenets. Talk of a unanimously agreed upon Islamic metaphysical framework is, therefore, not very helpful. Second, as Pierre Duhem had famously observed, "No metaphysical system suffices in constructing a physical [i.e., a scientific] theory."[25] Moreover, even assuming that it would be possible to produce such a framework, it does not mean that the science that will result within such framework will be "Islamic." This science must respond to reality for its validity. And no amount of pious Islamic intention can save a false scientific theory from being refuted by reality. During the Golden Age of Islam, Muslim scientists were able to produce many theories. Of these, those that were on the right track were incorporated into subsequent, more explanatory, theories and those that were not so, were simply rejected.

Hojjat al-Islam Mir-Baqeri, perhaps to avoid this particular rebuttal, has claimed that of all the sciences produced by Muslim scholars, only *fiqh* (Islamic jurisprudence), should be regarded as a genuine Islamic science and the rest should be termed as Muslim sciences. But even this move cannot help the project of rescuing Islamic science from failure. In the first place, *fiqh*, contrary to what Mir-Baqeri has suggested, is not a science at all, but only a technology: its aim is not to respond to Muslims' cognitive needs but to their practical concerns with regard to religious rules. Second, as was explained above, whatever is produced in whatever metaphysical framework can only be accepted as science if it genuinely explains some aspects of reality. It is the reality itself that remains the final arbiter for the validity of such an explanation and not the religious or nonreligious orientations or beliefs of those who subscribe to it.

Islamic Science and Technology: Once More

Following the above brief critical assessment of the views of some of the proponents of the project of the Islamization of knowledge, it can be asked what happens if Muslims succeed in creating a society where all the members are God-fearing Muslims who are all well-versed in Quranic and hadith studies and seek God's pleasure in all their activities including knowledge garnering and technology-building enterprises? Would it not be possible for these people to produce sciences and technologies that are radically different from what has been developed so far? In fact, this thought experiment could be greatly expanded and enlarged by asking what if the entire population of the Earth from the first moment when Homo sapiens emerged on the planet were such God-fearing human beings? Would it follow that human beings would develop radically different sciences and technologies from what are known so far?

Assuming that genuinely God-fearing societies of the sort described above consist of individuals with highest moral and cognitive standards, then the most that can be said about their scientific and technological achievements would be that the approach of these people toward the environment and other creatures including other fellow human beings could be expected to be more altruistic and less egoistic or greed-driven. They may, therefore, do less damage to the environment and the planet, or the universe, as a whole. However, given the fact that they are fallible human beings with limited cognitive capacities and are neither God nor angels nor infallible prophets or imams, then to comprehend reality, including God's words in His Book, they have no choice but to go through the long and tedious process of producing conjectures about reality's propensities and causal laws and testing their conjectures against the reality itself.

Since the reality before these pious people, is by definition, the same reality before us, these people would discover the same propensities and causal relations in reality. What these more pious individuals produce as scientific knowledge, in the final analysis, must be explanations of various aspects of reality. As such, these explanations are not, and cannot be, either Islamic or non-Islamic. They do not, and indeed should not, reflect or represent their discoverers' value systems. If they do so, they have failed in disclosing reality's secrets.

Moreover, what they discover, regardless of the notation and presentation system that they use for its communication, must explain the success of accepted mature scientific theories produced by more ordinary folk to date. These theories could easily be incorporated into their more-sophisticated theories as approximations and asymptotical cases.

Of course, they may discover secrets of reality faster than other, not so pious, human beings, of the sort that have appeared on the face of the Earth. They may even have the head start of knowing, like Adam (as the holy Quran says), the names (i.e., natures) of all beings,[26] but since they do not know the Unseen (*ghaib*),[27] they have no choice but to get their hands dirty through endless trials and errors to discover the secrets of reality.

What about technological achievements of these pious people? Could their technologies be regarded as Islamic or religious? Technologies, as we have already seen, are bearers of the values invested in them by their inventors or users. It is because of this feature that one can legitimately talk about a construct such as "Islamic democracy" or "Shari'a law"[28] However, even here the range and diversity of technologies that are genuinely Islamic or religious is not extensive, that is, they respond to specific and exclusive Islamic or religious noncognitive needs of believers and not their noncognitive needs as human beings. For example, in the past, the doors of houses in religious cities, in which opposite sexes would keep apart in the public, had two types of knockers, one a circular metal rod and the other a metal bar. The first one would produce a high-pitched noise while the other a low pitched. The first one was for communication between female visitors and female members of the household, while the second one was used by male visitors to announce themselves to the members of the household.

But given the common core that relates all human beings together and therefore produces universal values and needs, constructing exclusively religious technologies, though not impossible, would be rather difficult. In fact, given the common core of humanity, it can be claimed that producing exclusively religious technologies that could not be fine-tuned and used in nonreligious context would be almost impossible.[29]

What was said above is particularly pertinent to social and human sciences, which are the main target for some of the proponents of Islamization of knowledge. They maintain that the heart and soul of the project of Islamization of knowledge lies in these two fields. However, as was noted above, all social sciences and some of humanities are Janus-faced. They are both science and technology. As science, they cannot be Islamized in the same way that they cannot be Europanized or Indianized, since if what they discover about human nature is true, then it would be universally true. However, as technologies, it is in principle possible to add local values to these constructs to make them more effective tools for control and change in particular contexts.

* I would like to thank David Miller who made many useful comments on an earlier draft of this chapter.

END NOTES

1. Sayyid Jamal ad-Din Asad Abadi (al-Afghani). "Lecture on Teaching and Learning," in *Modernist Islam: 1840 – 1940*, edited by Charles Kurzman, 103–107 (New York: Oxford University Press, 2002).
2. Ibid. 106.
3. Seyyed Hossein Nasr. *The Need for a Sacred Science* (New York: State University of New York Press, 1993) 173; Seyyed Muhammad Naquib al-Attas, *A Commentary on the Hujjat al-Siddiq of Nur al-Din al-Raniri* (Kuala Lumpur: Ministry of Culture, 1970), 464–465.
4. Isma'il al-Faruqi and A. H. Abu Sulayman, *Islamization of Knowledge: General Principles and Work Plan* (Herndon, VA: IIIT, 1989), 57f, quoted in Leif Stenberg, *Islamization of Science* (Sweden: Lund University, 1996), 211.
5. Parvez Hoodbhoy, *Islam and Science* (London: Zed Books Ltd, 1991).
6. Mehdi Mir Baqeri. "Jonbesh-e Narm Afzāri va Chisti ye Ilm-e Islami" (The Conceptual Movement and the Nature of Islamic Science), *Siāsat-e Ruz* daily, November 10, 2003.
7. Medi Golshani. *Az Ilm-e Seculār tā Ilm-e Dini (From Secular Science to Religious Science)* (Tehran: Institute of Humanities and Cultural Studies, 2006).
8. Hossein Riyahi, "Religious Science: Feasibility and Structure," *Qabsāt* 8, nos. 76–77 (2004).
9. Javadi Amoli, *The Status of Reason within the Geometry of Religious Knowledge* (Qom: Asra, 2007). The title is apparently inspired by Ikhwan al-Sifa's (Brethren of Purity's) theory of knowledge.
10. Karl Mannheim, *Ideology and Utopia* (London: Routledge, 1929/1991), 261 and passim.
11. For a thorough treatment of critical rationalism see Karl Popper, *Conjectures and Refutations,* 5th ed. 1989 (London: Routledge, 1963); Popper, *Objective Knowledge* (Oxford: Oxford University Press, 1972); Popper, *The Myth of the Framework* (London: Routledge, 1994); David Miller, *Out Of Error: Further Essays on Critical Rationalism* (Aldershot: Ashgate, 2006); Miller, *Critical Rationalism* (Illinois: Open Court, 1994).
12. See, Karl Popper, "Three Views concerning Human Knowledge," in *Conjectures & Refutations* (London: Routledge, 2002); Ali Paya, "The Status of Intuition in Analytic Philosophy," in *Analytic Philosophy: Problems and Prospects,* revised ed. (Tehran: Tarh-e Nou, 2011).
13. Pierre Duhem. *The Aim and Structure of Physical Theory* (Princeton, NJ: Princeton University Press, 1954/1991), 187.
14. Joseph Agassi, *Technology: Philosophical and Social Aspects* (Dordrecht: Reidel, 1985). It must be stressed that what is said in the text does not entail that technological progress does not benefit from past experiences and a gradual process of improving upon earlier technologies. Since technologies ought to respond to the needs of users in specific contexts, adjusting them to these contexts usually requires some personal touch, finesse, and adeptness that, contrary to scientific knowledge, are not part of an objective and detached world. Language is a good case in point. When the last speaker of a language dies, the way it is spoken by its native speakers is lost forever. It should also be

noted that although instruction manuals are used to instruct users of technologies of the know-how of operating the machines and instruments, users' levels of mastery in dealing with machines and technologies are not equal. Some are far better than others in this respect. It is this extra skill and ability that is user-specific and not transferable by objective means.

15. Of course, pragmatic considerations, in the final analysis, rely on the notion of truth for their credibility: an effective instrument is the one that remains true to its design, assuming that the design itself is a correct one, free from errors of judgment.

16. Martin Heidegger, "The Question Concerning Technology," *Basic Writings* (New York: HarperCollins Publishers, 1993).

17. It must be emphasized that fundamental laws governing human nature are not psychologistic of the sort advocated by John Stuart Mill and his psychologistic school of sociology that was, as Popper has observed, joined later by many psychoanalysts (Karl Popper, *Open Society and Its Enemies* (Princeton, NJ: Princeton University Press, 1971) ch. 14). These laws are not reducible to the psychological laws of human nature, but "emergent laws" that belong to more complex (living) systems. For a discussion of emergent biological laws that govern human behaviour, see Paul Davies. "Emergent Biological Principles and the Computational Properties of the Universe," *Complexity* 10, no. 2 (2004): 11– 15; H. J. Morowitz, *The Emergence of Everything* (Oxford: Oxford University Press, 2002).

18. Some writers maintain that a further distinction, between phenomenological and empirical laws, is needed. See for example, Samuel Schindler. ' "Bottom-up' Corrections and the Phenomenologically-driven View," paper presented at *10th Annual Oxford Philosophy Graduate Conference*, University of Oxford, November 2006. A handout of the paper can be obtained at http://www.samuelschindler.org/talks.html. However, it seems such an analytical hair-splitting does not make substantial difference to the issue concerning fundamental and nonfundamental laws.

19. Michel Le Bellac and Patricia de Forcrand-Millard, *Quantum Physics* (Cambridge: Cambridge University Press, 2006), 11. The authors explain that most phenomenological laws in the field of physical sciences are nothing but a first term of a Taylor series.

20. Nicholas Maxwell. "The Need for a Revolution in the Philosophy of Science," *Journal for General Philosophy of Science* 33 (2002): 381–408; Duhem, *The Aim and Structure of Physical Theory*, 191–195.

21. Nancy Cartwright. *How the Laws of Physics Lie* (Oxford: Oxford University Press, 1983), 107–110.

22. Ali Paya, "Do the Fundamental Laws of Physics Furnish Us with a Faithful Picture of Reality?" in *Analytic Philosophy and Philosophical Analysis: Problems, Perspectives, and Applications* (Tehran: Tarh-e Nou, 2011).

23. Hossein Bostan, *Gāmi be Su-ye Ilm-e Dini: Sāakhtār va Imkāan-e Tajrobi-ye Ilm Dini (A Step Towards Religious Science: The Structure and Empirical Feasibility of Religious Science)* (Iran: Institute of Hawzeh va Daneshgah, 2005).

24. Pragmatism is a broad church. Various writers have defined it in different ways. But generally speaking it can be captured in the following formulation: "truth

of a belief consists in its working." Gerald Vision, *Modern Anti-realism and Manufactured Truth* (London: Routledge, 1988), 103. "Working," however, means different things to different people, hence relativism. Moreover, such a view of truth reduces all knowledge inquiries to pursuit of what "works"; it equates science with technology. Furthermore, contrary to what the pragmatists claim, "working" does not constitute truth: something may *work* for a subject without having any bearing on truth about the objective reality. The naked emperor, in Hans Christian Anderson's story, was over the moon, since he believed that he had obtained the best cloth on earth, but the truth was what that young boy in the crowd shouted. For further details see, Gerald Vision (1988). Ali Paya, *Analytic Philosophy and Philosophical Analysis: Problems, Perspectives, and Applications* (Tehran: Tarh-e Nou, 2011).

25. Duhem, *The Aim and Structure of Physical Theory*, 16.
26. "And He Taught Adam the names (the nature of all things)," The Quran, 2:31.
27. Even the Prophet Mohammad emphasized that he did not know the Unseen, otherwise he could have multiplied all good and avoided all evil: "Say: 'I have no power over any good or harm to myself except as God willeth. If I had knowledge of the unseen, I should have multiplied all good, and no evil should have touched me." The Quran, 7:118.
28. For the possibility of developing Islamic models of democracy see Ali Paya, "*Islamic Democracy:* A Valid Concept or an Oxymoron?" in *Iraq, Democracy, and the Future of the Muslim World*, edited by Ali Paya and John Esposito (London: Routledge, 2010).
29. It is indeed evident that the invention of two types of door knockers could be adapted by non-Muslims to nonreligious needs. Before the days of mobiles, some American families used to have two numbers in the telephone directory, one of them marked "Teenagers." (I owe this example to David Miller).

9

THE ROLE OF INTERCULTURAL
DIALOGUE IN THE RISE OF
MODERN SCIENCE

Anjam Khursheed

INTRODUCTION

The fact that many developments critical to the rise of modern science took place in Asia during the period known traditionally as the "Dark Ages" or the "Medieval Period" is now becoming better known. However, their description is usually found within the framework of single cultural heritages, where Chinese, Indian, or Islamic civilizational contributions to science are described separately.[1]

Recently, there have been attempts to take into account the impact of more than one Asian culture,[2] and more importantly, there is an effort to view the whole process from a multicultural perspective, where the rise of modern science is better understood as something that emerged from the dialogue of different cultures.[3] The aim of this chapter is to explore the multicultural perspective further. First, some areas of intercultural exchange within Asia will be highlighted. Second, some misconceptions about the dialogue that occurred between Asian civilizations and Western Europe will be clarified. In particular, the chapter will challenge the widespread belief that it was the reemergence of ancient Greek philosophy in the late medieval period that brought about an intellectual recovery in Europe. The view put forward in the following pages is that ancient Greek influence was one among many cultural influences, and it was primarily a dialogue with the living civilizations of Asia that brought Latin speaking Europe out of its Dark Ages.

SCIENTIFIC AND TECHNOLOGICAL
DIALOGUE IN ASIA

Around a millennium ago, the continent of Asia led the world in science and technology, and intercultural dialogue was an essential part of this achievement. The communications channels linking different cultures together were numerous and diverse, but foremost amongst them, was the great uniting influences of Buddhism and Islam. Buddhism reached China in the early part of the first millennium CE, and in succeeding centuries, became an integral part of Chinese culture. Not only did Buddhism link China with India, but it opened up channels of communication throughout East Asia, both overland and by sea. One of the most traveled links was the one shared by both merchant and monk, the silk/spice trade route via the Indian Ocean, which was also important for the sharing of scientific and technological ideas. With the rise of Islam in the second half of the millennium, new routes of communication were created and older ones greatly strengthened: overland silk routes linking China to Western Asia (Iran, Iraq, and Syria), as well as enlarged spice sea routes that now linked China to the Middle East, Africa, and the Iberian peninsula. Islam in its early history was famed for its openness in acquiring knowledge from foreign cultures, and like Buddhism, it provided a positive means of communication through which Asian cultures could exchange technological and scientific ideas.

Buddhism achieved its integration of cultures primarily on an individual level. It spread peacefully; its method of persuasion was through conversation, and its activities were largely confined to the quiet of monastic life. Islam, on the other hand, spread mainly through conquests and political unions. The large territories that fell under its rule were subsequently united through a spectacular array of multilingual, multiethnic, and multireligious centers of learning and trade.

Until relatively recently, the great Chinese contribution to the development of science and technology was not widely appreciated. Only after the Second World War, largely through the seminal work of the English-scientist/historian Joseph Needham, were the great achievements of Chinese science better recognized.[4] Three Chinese inventions in particular, printing, gunpowder, and the magnetic compass, are conspicuous of the great scientific legacy that China bequeathed to Europe. These inventions transformed medieval Europe and prepared it for the modern era. By the seventeenth century, many European scientists and philosophers were acknowledging their debt to these inventions, although curiously, they did not know where these inventions originated. The well-known seventeenth-century English spokesman of modern science, Sir Francis Bacon, refers to the impact of these inventions in the following way:

"It is well to observe [said Lord Verulam] the force and virtue and consequences of discoveries. These are to be seen nowhere more conspicuously
than in those three which were unknown to the ancients, and which the
origin, though recent, is obscure and inglorious; namely, printing, gunpowder, and the magnet. For these three have changed the whole face
and state of things throughout the world, the first in literature, the second in warfare, the third in navigation; whence have followed innumer-
able changes; in so much that no empire, no sect, no star, seems to have
exerted greater power and influence in human affairs than these mechanical discoveries."[5]

Bacon's clear testimony that printing, gunpowder, and the magnetic
compass were among the most powerful scientific discoveries of his
day is important, as is his ignorance concerning their origins. It is
difficult to exaggerate the great social impact these inventions had
on European life in the late medieval period. The Reformation,
Renaissance, and the rise of capitalism all made use of printing, and
the democratic form of education that it made possible. Gunpowder
weapons brought about a more democratic form of warfare, and led
to the end of Western military aristocratic feudalism. The castle was
no longer the most secure fighting post. Both at sea and land, gunpowder made war more mobile. The magnetic compass opened up a
new era of navigational science for Europe in the fifteenth century
and took Europeans around the world, eventually to discover America.
The great social transforming effect of such expeditions gave rise to
Western colonies and brought the world much closer than it had ever
been before. Also, no less important, is the role that magnetic science
played in the development of modern science. The compass, the form
of an accurate detector of magnetic fields, played a crucial part in finding a link between electric currents and magnetic fields, which subsequently led to the inventions of the electromagnet, electric motors,
generators, telephones, and so many other electrical inventions that
have transformed the modern world. The study of magnetism provided parallels by which the gravitational force was understood, and
this body of science finally evolved into field theory. Field theory is an
important branch of physics by which we understand electromagnetic
and gravitational phenomena today.

It is extraordinary, given the far-reaching effects of printing, gunpowder, and the magnetic compass on European society that their origins
were to remain "obscure." What is even more surprising is that for most
Europeans, the origins of these inventions remained a mystery right up
to the twentieth century. Indeed, Needham cites two occurrences of it in
books written by well-known European scholars, one that was published
in 1920, and the other published in 1962. Needham refers to it as an
attitude of "invincible ignorance."[6]

The Chinese wrote of magnetic effects as early as 83 CE, in relation to divination devices called *Shih*. The earliest description of a magnetic compass dates from the ninth century CE. A small wooden piece of loadstone is embedded in the body of a wooden fish that is floated in water. The fish has a little needle projecting from it that points south. An interesting feature about this arrangement was that the compass-fish was magnetized by being heated to red heat while held in the north-south position in the Earth's magnetic field. This shows that the Chinese knew about remnant magnetism. There is also evidence that they knew of magnetic declination (the fact that compasses do not point exactly north-south and that the difference varies with time). The Chinese used the magnetic compass in navigation around the tenth century CE. They achieved all these things long before Europeans even knew of magnetic polarity. The earliest mention of magnetic polarity in Europe dates from 1180 CE. This puts China ahead of Europe by at least two to three centuries. However, Arabic commentators certainly knew of the Chinese fish-compass, and often referred to it. It is thus quite likely that information about the compass and magnetic effects came to Europe via the Islamic civilization.

The development of gunpowder shows signs of being part of the wider Chinese-Indian cross-cultural dialogue. In 664 CE, an Indian was able to identify soils in China that contained saltpeter, and demonstrate the purple flame created when it is put in a flame. Later-Chinese studies also show evidence of an Indian influence, which seems to have been the starting point for investigations that led to the first recipes for gunpowder.[7] The first mention of an explosive chemical mix resembling gunpowder occurs in the ninth century CE. In 919 CE, the "fire drug," as it was known, was used as ignition for a flamethrower. By 1000 CE, it was packed into bomb and grenade form. The first composition formula appeared in 1044 CE, some three centuries earlier than references to gunpowder composition in Europe. These early gunpowder devices were more like rockets, and did not consist of the much later destructive explosive mix. In fact, rocket devices in China date from around the eleventh century. Gunpowder weaponry in Europe started to develop in the early fourteenth century.

Printing was also a technology that developed within the wider Chinese-Indian cross-cultural dialogue. A Chinese monk while visiting India reported the impressing of the Buddha's image on silk or paper. This related to the Indian technique of printing patterns on cloth from carved wooden blocks, but it was never developed for printing books. Gradually, the idea of carving whole pages of writing on wood and impressing them on paper arose in China.[8] Block printing in China developed in the ninth century CE. The earliest printed book is a Buddhist text (The Diamond Sutra), which dates from 868 CE. In 932 CE, the complete printed edition of the Classical Books of Confucius was commissioned. Moveable type was developed in the eleventh century, even though a separate piece of type was needed for each one of the thousands of Chinese characters. Moveable type was introduced into Europe some four hundred years later, when Gutenberg printed his Latin Bible in 1456 CE.

There was much more to the Chinese technological heritage than the development of gunpowder, the magnetic compass, and printing. Other contributions of China, according to Needham, include, the efficient equine harness, the technology of iron and steel, the invention of the mechanical clock, the development of basic engineering devices such as the driving-belt, the chain-drive, the standard method of converting rotary to rectilinear motion, segmental arch bridges, nautical techniques such as the sternpost rudder, the seismograph, and deep drilling, to mention only a few. The list is considerable.

Although the Chinese contribution in technology is impressive, the idea that all these inventions came only from China is misleading. There are several instances of Needham's Chinese inventions being shared by other cultures. Take for instance the invention of the mechanical clock, whose development Needham claims to be mainly Chinese.[9]

A clock driven by a waterwheel is described by the Chinese scientist, Su Sung in Khaifêng, the capital of the Northern Sung Dynasty in 1092 CE. This clock used a linkwork escapement mechanism that made it accurate. Clocks in Europe did not arrive at the same accuracy of Su Sung's clock until the seventeenth century. The first origins of clock making in China started around 725 CE, when a Tantric Buddhist, I-Hsing, and an engineer, Liang Ling-Tsan, built a waterwheel clock for an imperial court of higher learning.

The development of the clock had of course, a profound influence on the craft tradition that accompanied modern science. Needham speculates that the Chinese waterwheel linkwork clock was known in thirteenth century Europe, or that there was at least the knowledge that the problem of mechanical timekeeping had been solved. A channel of communication between China and Europe was certainly created through the thirteenth-century Mongol invasion of Eastern Europe, and it is likely that knowledge of Chinese clocks reached Western Europe at that time through countries like present-day Russia. But knowledge of clocks, as with many other scientific developments of Asia, reached Europe through the Islamic civilization. In fact, waterwheel driven clocks were designed and constructed in the Islamic world precisely around the same time that they appeared in China, and there is disagreement among historians whether medieval Muslims knew of Chinese clocks or vice versa. Historian Arnold Pacey cites the Muslim contribution in the following way:

> Among Islamic books studied or translated at Toledo, there were several which discussed mechanical devices including astronomical and several types of water clocks. One author who wrote on this subject was al-Muradi,

and he illustrated elaborate gear trains, some with epicyclic and segmental gears. It is particularly interesting to note that he was working at almost exactly the time Su Sung was building his great clock in China. Indeed, one of al-Muradi's designs was for a clock driven by a water-wheel like Su Sung's, but no connection between the two seems likely. A more relevant connection is with two water-clocks of rather simpler design which were operating at Toledo in the 1080s.[10]

The relationship between clocks in the Islamic and Chinese civilizations is not well documented. It may be that Muslims learned of Chinese clocks, and made developments of their own, or, it could be that there was much more exchange of technology than we suppose, and that waterwheel clocks was a shared invention. A plausible line of development of European weight-driven clocks being based upon prior mercury clocks made in Islamic Spain has recently been proposed.[11] The mercury acted as a kind of hydraulic escapement, and a train of segmental gears was used to transmit high torque. Apart from the innovation of having a mechanical form of escapement, later added in European weight-driven clocks, all the essential elements were present in the Islamic mercury water clock from which the later weight-driven European clock could have developed. The earliest description of a European clock of this kind is a clock at the Benedictine monastery of Santa Maria de Ripoll at the foot of the Pyrenees, dated sometime after 1271 CE. Apart from the mechanical escapement mechanism, the style of the clock resembles the mercury clocks that were known to exist in Islamic Spain. It is, therefore, plausible to think that the European weight-driven clock evolved from prior mercury clocks in Islamic Spain, incorporating the added innovation of a mechanical escapement.

Evidence of there being shared scientific influence between Islamic and Chinese civilizations is also apparent in astronomy. Waterwheel clocks in China were used to drive astronomical devices called "armillary spheres," which mapped the coordinates of stars on the celestial sphere. Needham cites the Chinese astronomer Kuo Shou-Ching for the development of a bronze armillary sphere in 1275 CE at Peking, which he takes to be the forerunner of the equatorial mounting of the modern telescope. Modern astronomy uses the Chinese system of equatorial coordinates to measure a star's position on the celestial sphere, rather than the ecliptic coordinates of the Greeks or the altazimuth measurements of the Arabs. But Needham himself explains that Kuo Shou-Ching arrived at his invention by modifying the "torqutum," a kind of computing machine for performing transformations between coordinate systems, which was first designed by the Spanish Muslim Jabir ibn Aflah, and that it was introduced into China by the scientific mission of Jamal al-Din in 1267 CE.[12] This is an instance where a Chinese invention was inspired by Islamic science. Moreover, the explicit mention of there being an Islamic "scientific mission" in China does seem to suggest that Muslims at least, were looking for scientific collaboration.

Needham records an instance of the Chinese setting out to learn from the Arabs. The famous Muslim physician Al-Razi records the visit of a Chinese scholar to his home in Baghdad in the tenth century. Al-Razi wrote of how his Chinese guest stayed there learning Arabic in six months, and then translated the works of Galen into Chinese, after which he left to return to China. Chinese and Islamic scientists collaborated under Mongol rule. The Mongols instigated an astronomical observatory in the thirteenth century at Maraghah (south of Tabriz in Iran) under the care of the famous Nasir al-Din al-Tusi. The Mongols arranged for a collaboration of Chinese astronomers (Fu Meng-Chi) with Muslim astronomers that came from Spain (al-Maghribi, al-Andalusi). Later, a similar astronomical observatory was set up in Samarkand in Central Asia.[13] The Islamic-Chinese contact, although officially sanctioned and encouraged by the Mongols in the thirteenth and fourteenth centuries, had certainly taken place earlier, in the Sung and Tang dynasties. Islam's overland entry in China is likely to have occurred in the mid-seventh century.[14] During the Tang dynasty, Arab traders were frequently in Canton. Mosques started to appear in China before the turn of the first millennium. All this suggests that there was ample opportunity for scientific interchange.

The technological dialogue between countries under Islamic rule and China was generally mutually beneficial. For instance, silk and expertise on how to make it supplied from China led to the development of the textile industry in Western Asia, while, innovations such as the wind-mill came from Iran, and there was much shared experience in large-scale hydraulic systems, such as irrigation of land and the building of dams. Islam in China of course, reached its height in the Yuan Dynasty of the thirteenth century, during which the Mongolian emperor Kublai Khan recruited many Persian and other foreign administrators into the Chinese civil service, and sent engineers to Iran on joint projects. Many high-ranking officials within the Chinese Navy were Muslims, such as the famous admiral Cheng Ho, who led the "Treasure Fleet" out to explore the Indian and western Pacific oceans. Cheng Ho's father had made a pilgrimage to Mecca.

On the Indian-Chinese exchange, there are Chinese references in 636 CE to many Brahmin books on astronomy, medicine, and mathematics that are now lost. Around the same period, there are Chinese references to Indian knowledge of mineral acids.[15] There are records of Buddhist activity in building and repair of bridges along routes connecting China to India, and large bronze statues of the Buddha have been found as early as 734 CE, which must have involved the use of metal furnaces capable of melting about 1 ton of metal.[16] Already mentioned, are the joint developments of gunpowder and printing, and the oldest printed book, a copy of the Diamond Sutra, the most sacred of all Chinese Buddhist sutras. By the twelfth century CE, Buddhist mathematical textbooks in China were very common. A twelfth-century Confucian scholar cited by Needham states, "Nowadays even children learn mathematics from Buddhist textbooks which deal with the counting of infinite numbers of sand-grains"[17]

In mathematics, there are many unanswered questions in the exchange. Although Needham does refer to translations of various Indian mathematical works: a table of sines by the Indian mathematician Aryabhata,[18] the work of I-Xing, a Buddhist monk credited for calculating the tangent table, Qin Jiushao being the first to introduce the zero symbol into Chinese mathematics (which was not subsequently adopted),[19] for the most part, Needham describes Chinese mathematics as an independent development. However, there are many similar mathematical innovations in algebra and the solution of equations that also occurred in the Islamic civilization. The historian of mathematics G. G. Joseph notes that the Chinese way of solving numerical equations of higher order may have influenced Arab mathematics, conversely, some details of Arab trigonometry and algebra are likely to have reached China.[20] Knowing the many channels of communication and collaboration that existed between China and Islamic countries, one suspects that many of these innovations also arose out of cross-cultural exchange.

Needham acknowledges that the interchange of scientific information between China and other Asian civilizations such as the Indian and Islamic ones requires further research.[21] But one important point emerges from the scanty information that already exists. It is not possible to study Chinese science in isolation, without reference to other Asian cultures. In the period 600–1200 CE, during the so-called Dark Age of Europe, three Asian civilizations reached their height: Chinese, Indian, and Islamic, and they had considerable contact with one another. During the Tang dynasty, which was when Buddhism in China was at its height, some of China's most-enduring scientific discoveries such as gunpowder and the mariner's compass were made. Reference has already been made to specific inventions such as bridge-building, and printing, which had direct Buddhist involvement. The great scientific achievements of the Sung dynasty in China roughly correlate with similar developments in the Islamic civilization. They include improvements in clock design, construction of mechanical devices for astronomy, and the use of algebra. The very best of Chinese science, which occurred in the Tang and Sung dynasties, took place when China was open to Buddhist and Islamic cultures. They were periods when China's links with other countries were strong, involving connections through Central Asia to Islamic countries in the West, across the mountains south to India, eastward to Japan, and down by sea around Southeast Asia.

THE EAST-WEST MEDIEVAL DIALOGUE

The gradual diffusion of science and technology from Asian civilizations to Latin-speaking Western Europe from the turn of the first millennium CE was a process that continued on for more than six hundred years. It proceeded on many different levels, ranging from the work of scholars to the transfer of technology through trade and warfare. The city-state of Venice and its sizeable territories spread around the Mediterranean and the Levant (such as Syria) became the focus of East-West trade in the thirteenth/fourteenth centuries. Another channel of communication was opened up by the thirteenth-century Mongolian conquest of Eastern Europe. Later, in the fifteenth century, the famous Portuguese expeditions to India and around Southeast Asia opened another line of communication. The historian Arnold Pacey cites some examples of the kind of technology transfer that took place along these East-West trading routes.[22] He cites skilled Chinese introducing the method of papermaking at Samarqand and Baghdad, and Islamic workers taking the technique to Spain. This must have been the most common way in which Chinese technology made its way into Christian Western Europe, and this also illustrates the vital role played by the Islamic civilization in the East-West dialogue. Another example of this is silk making. In 1050 CE, there was a silk industry in areas ruled from Constantinople and parts of the Islamic world, including Tunisia, Sicily, and Granada (Spain), but no record of it in the Christian Western parts of Europe. However, just before 1300 CE, silk manufacture started at Lucca, in Northern Italy, and it appears that knowledge of relevant techniques had been obtained from Sicily. In this way, silk-making technology from China had been transferred to the West. It is interesting to note that the famous glass-making tradition of Venice was originally transferred from Syria in the thirteenth century, and is a part of the wider transfer of considerable chemical knowledge from the Islamic world. The transfer of technology was of course, a two-way creative exchange. Pacey describes how the transfer of gunpowder recipes and simple "fire-lance" weapons from China stimulated the invention of the cannon in Europe.[23]

Despite there being much evidence on scientific and technological transfer from Asia into Christian Western Europe in the centuries preceding the European Renaissance and Scientific Revolution, the importance of this is not usually acknowledged within Western European historical accounts of modern science. In fact, as illustrated by the quote made by Francis Bacon at the beginning of this chapter, the origins of gunpowder, printing, and the compass were not well-known. The following discussion will examine certain ideological elements within medieval Christian Western Europe that acted as barriers to East-West dialogue.

There is a standard misconception about the growth of science—a popular myth that has come to dominate the Western European mind ever since the close of the Middle Ages. According to this myth, science originated in ancient Greece. Then, following the collapse of Rome at the end of the fifth century CE to the beginning of the twelfth century, there was a "Dark Age," during which most of Europe was overrun by Germanic tribes whose invasions stunted cultural growth, embroiled European people in wars, and feuds whose impact lasted well over seven hundred years. Meanwhile, Greek writings in philosophy were preserved by the Arabs. During this time, scientific progress came to a halt. When ancient Greek writings were transmitted back to Europe from the twelfth century onward, they started to cause an intellectual revival, particularly through the works of Aristotle in the thirteenth century, and later through Plato's writings in the fifteenth century, and it was this revival of learning that culminated into the European Renaissance out of which modern science grew.[24],[25]

How this Eurocentric Dark Age Greco myth came to dominate the Western European mind for many centuries is a complex question. No doubt the propaganda associated with the crusades provided much of the initial sociopolitical impetus to it, but that does not explain its duration up to the twentieth century. Perhaps, it has more to do with the fact that political rivalry with Muslim governments continued on from the Middle Ages in the form of wars with the Ottoman Empire, and did not abate until the Young Turks revolution of 1908. From a political standpoint, it seems that Muslim and Christian governments have been at war for around a millennium.

There are many obvious weaknesses with the Eurocentric Dark Age Greco myth, which usually cites recovery of the writings of Aristotle and Plato as being among the important elements responsible for Europe emerging out of its medieval cultural obscurity. The myth gives the impression that Greek learning in the centuries following the sack of Rome in the fifth century had been lost in Europe, or at least, knowledge of it had been so greatly reduced, that it barely existed, but this is not what the historical record shows. Before the collapse of Rome, Constantine had moved the seat of the Roman Empire to a small town called "Byzantium" (in modern-day Turkey), later to be known as Constantinople. The Byzantine Empire flourished for more than 1000 years, before it was captured by the Turks in the fifteenth century. The language of the Byzantine Empire was Greek, and in fact, Greek culture in Europe was very much alive throughout the Middle Ages. Not only was it alive, the Byzantine Empire was one of the most powerful economic, cultural, and military forces in Europe; it extended over most of Eastern Europe and to territories in the north, including Russia. There was, in fact, ample access to Greek literature and culture in Europe throughout the entire period of the Middle Ages; the problem has more to do with the fact that Europe was irreconcilably divided.

The former Western Roman Empire, centered at Rome, developed into a Latin-speaking Christian Catholic Europe, while the eastern part, centered at Constantinople, developed into a Greek-speaking Christian Orthodox Europe (Byzantine Empire). Western and Eastern Europe were embroiled in bitter religious and political conflicts. The power struggle was not just a political one, the head of Western Europe, the pope of Rome, and the head of Eastern Europe, the patriarch of Constantinople, were battling over theological leadership. In the eleventh century, they simultaneously excommunicated one another (known as the Great Schism), leaving Eastern and Western Europe deeply divided into two irreconcilably separate domains. Their political-theological disputes continued, even while attempting to combine forces to combat the Muslims in Jerusalem—the "Holy Wars" of the crusades. In 1204 CE, Western European crusaders captured and sacked Constantinople, imposed a Latin Patriarch, and destroyed its library and many of its valuable religious artifacts. There were, of course, many attempts at reconciliation, but Western and Eastern Europe remained bitterly divided throughout the Middle Ages.

The great cultural division of Europe into a Latin-speaking part and a Greek part, may explain why later (Western European) historians refer to knowledge of Greek culture as being lost to Europe, even though it was actually very much alive. It was just that for a long time (around 1000 years), Greek was the language of the enemy. For the Latin-speaking world, there was in fact, ample opportunity to acquire and read Plato and Aristotle throughout the Middle Ages; what was lacking was not the opportunity, but the desire. During the Dark Ages (early Middle Ages), Latin-speaking Europeans did not look upon Greek culture as something that belonged to them, but as something that belonged to a rival cultural tradition.

It is interesting to note that the Byzantine Empire had the benefit of much greater political and cultural stability than the Latin West; it had centers of learning in Constantinople, Alexandria, and Athens. It had its own philosophers, such as John Philoponus, who taught in Alexandria. Pliloponus opposed various scientific aspects in Aristole's writings. He, for instance, rejected Aristotle's account of projectile motion, and called for consistency between the Aristotle's terrestrial and celestial theories. In all of this, of course, he predates many concerns that were to occupy Copernicus and Galileo. Apart from commentaries on Aristotle, there were commentaries on the writings of famous Hellenistic philosophers and scientists, such as Ptolemy, Hero of Alexandria, Archimedes, and Appollonius of Perga.[26] All this suggests that the Dark Age myth does not even apply to the European continent—it is perhaps more accurately applied to Western Europe, rather than Europe as a whole. Even here, it would be inaccurate, since science and philosophy flourished in Islamic Spain and Sicily; so it is perhaps only applicable to Latin-speaking Western Europe.

Recognition of the deep cultural divide in Europe in the Dark Ages might explain why transmission of Greek culture to Western Europe first came primarily through contacts with the Islamic civilization. There were, in the early part of the twelfth century, at least four centers of exchange where Western European scholars could learn more about science. A small number of Western scholars traveled to Syria in the wake of the first crusade, settling in cities such as Antioch, where they found and translated books from Arabic. Second, there was Constantinople, already mentioned, which despite having good links with Northern Italian cities such as Venice and Pisa, did not attract the attention of many scholars. There were, of course, some scholars, such as James of Venice, who translated the whole corpus of Aristotle's writings of Logic, but such scholars were relatively few, and their work did not seem to attract much attention. Third, there was the theater of interchange in Sicily, under its Norman monarchs. Sicily had, of course, been under Islamic rule, and it became an important center for translation, particularly under King Roger II. However, by far the biggest source of learning and translation of Greek ancient documents occurred in Islamic Spain. In terms of the sheer volume of translations, the number of scholars involved, and the influence that their work subsequently had on Latin-speaking Europe. Spain around the turn of the first millennium was the main center of exchange through which scholars in Western Europe became aware of the scientific achievement of Asian civilizations. The fact that more direct acquisition of ancient Greek culture was available from Constantinople is something that recent historians are also beginning to note.[27]

Around a millennium ago, broadly speaking, in terms of dialogue between civilizations, there was a high point in Asia, and a low point in Europe. In contrast to the positive communication channels between cultures created by Buddhism and early Islam in Asia, known for their openness, inclusiveness, and promotion of foreign learning, Europe was characterized by bitter religious-political rivalries, where religious exclusivism intermingled with power politics. There the Holy Wars raged between Islam and Christianity from the beginning of the twelfth century to the middle of the thirteenth century. For Christians, the religious frenzy of seven crusades, starting from the declaration of the first crusade in 1095 to the recapture of Jerusalem in 1249, involved not only the killing of Muslim infidels in the Holy Land, but also spread to large-scale massacres of Jews throughout Europe. For the papacy, the crusades were only one of a series of wars. Others had already been waged against the Eastern Orthodox Church and yet others were being fought against the Roman emperor, and the kings of France and England. In the late fourteenth century, there were three popes vying for power, a conflict known as the "Western Schism." The kind of ideological exclusiveness that fueled many of these conflicts was an all-embracing one, not just religious, but political, social, and cultural. It is, therefore, perhaps not surprising that a Eurocentric view on the history of science emerged from the late Middle Age period.

What was it that reversed this situation in the space of say four hundred to six hundred years? According to most accounts in the history of science, the first part of the intellectual recovery in Western Europe occurred when "scholasticism," forged an alliance between Aristotle's philosophy and Christian theology. Aristotle's vast array of empirical studies on Nature, his use of rational and logical principles to classify and catalog his observations are thought to have loosened the hold of Augustine's "world-denying" theology over the medieval Christian mind.[28] It is undeniably true that Aristotle's writings exercised a strong influence over thirteenth-century scholastic minds; he was the most respected of all philosophers and was simply known as "The Philosopher," just as St Paul was referred to as "The Apostle." Aristotle's philosophy came to dominate the study of subjects outside theology in the curriculum of Europe's first universities, such as those at Bologna, Paris, and Oxford, which started to develop in the thirteenth century. But having recognized this fact, it may be pertinent to ask what part of Aristotle's philosophy was it that inspired progress toward modern science? In fact, taking a broader perspective, it may be plausibly argued that Aristotle's philosophy was more of an obstacle than an inspiration to the development of modern science. The seventeenth-century pioneers of modern science, in one way or another, disproved many of Aristotle's scientific propositions about the natural world. Whether it be in relation to a heavier mass falling faster than a lighter one, or of the nonexistence of a vacuum, or the path of a projectile being rectangular in shape, most of Aristotle's theories were overturned by the rise of modern science. The very method that Aristotle used was also rejected. Aristotle's scientific method was founded on making observations and then framing theories, but the seventeenth-century pioneers of modern science found that observations by themselves were not enough; they stressed the need to carry out controlled experiments and make scientific instruments to make empirical measurements, as well as the need to experimentally test theories—something that was completely missing in Aristotle's philosophy.

The making of scientific instruments and the carrying out of experiments to test theories, in ways that went beyond ancient Greek science, are now widely recognized to have occurred in the Islamic civilization. Scientific instruments such as the astrolabe, armillary sphere, abacus, and water clocks were quite commonly found in many countries under Islamic rule around a millennium ago, and they certainly became available to Western Europe in places such as Spain in the eleventh century, as a result of its "Reconquest." The contributions of Islamic science in the field of optics is a good illustration that significant aspects to the "modern scientific method" were developed in the Middle Ages, within the Islamic civilization, and not only in the sixteenth and seventeenth centuries in Europe. The most famous Islamic mathematician-physicist in this field was the eleventh-century born Ibn-al-Haitham, known as Alhazen in the Latin-speaking world. Alhazen, building on the work carried out by other Arabic scholars (such as Al-Kindi and Al-Nayrizi), made discoveries that went well beyond the ancient Greek studies in this area, and at the same time tested his mathematical theories with a systematic method of experimentation. He inspired men such as the English scientist-philosopher Roger Bacon in the thirteenth century to carry out similar scientific experiments, and also drew praise from men such as the fifteenth-century Italian Renaissance artist-engineer Leonardo da Vinci, and the seventeenth-century German astronomer Johan Kepler for his contributions to science. Alhazen's theories in optics laid the foundation of the principles of perspective in the visual arts and his authority in the field of optics lasted well into the seventeenth century where it was further developed by Isaac Newton.

The Greeks had assumed that rays of light originated from the eye, whereas Alhazen established the fact that they leave the luminous object observed by the eye. He gave an accurate and detailed account of the operation of the eye and how it functioned. He also generalized the laws of the reflection of light from plane mirror surfaces to concave and parabolic ones, and related the laws of refraction to the solid density in which the light was deflected. This latter observation was extended to describe the influence of the atmosphere on the light from luminous objects in the sky. He modeled problems in optics by mathematics. One such problem led to him solving a fourth degree algebraic equation by geometrical means and became known as the "Alhazen Problem" in the Latin-speaking world. Performing experiments on his ideas, he constructed ingenious mechanical apparatus, such as the steel refractor lenses. In all this work, Alhazen displayed a degree of mathematical and experimental rigor that had not existed in ancient science and that greatly contributed to the development of modern science—six centuries before the emergence of the Scientific Revolution of the seventeenth century. Historian D. R. Hill has summarized the contribution of Alhazen in the following way:

He was quite prepared to modify or even reject a hypothesis if it conflicted with experimental results. He took great care in the construction and the assembly of the equipment for his experiments, and made the radical innovation of including dimensions as an integral part of his specifications—an essential element in any serious experiment. Even so, the practice is not to be found in earlier works on optics. Ibn al-Haytham's experiments, apart from verifying the conditions for vision were also applied to problems of reflection and refraction. One of the results of his methodology was the development of precision instrumentation. Although not alone in this field—astronomers and surveyors, for example, made notable advances in the construction of accurate instruments—Ibn al-Haytham's contribution undoubtedly led to significant improvements in instrument design.[29]

The importance of making precision scientific instruments for the purpose of carrying out experiments in the rise of modern science cannot be overemphasized. Galileo and William Gilbert are perhaps the best examples of the first Western European scientists who used this approach. Gilbert performed many experiments in electricity and magnetism, such as his famous spherical loadstone experiment, by which he concluded the Earth to be a giant magnet. Galileo, of course, rolled balls down an incline to show that two balls of differing mass move with the same acceleration, and he made his own telescope, making improvements to the original design. He even ground his own lenses.

What specific mathematical theories did Plato contribute to the development of modern science? It is true that there arose a strong mathematical tradition amongst fifteenth-century humanists in Northern Italy, and it is true that they studied and celebrated Plato's philosophy—we know that they created an academy dedicated to the study of Plato in Florence. While it is true that the humanist movement did place greater emphasis on mathematics rather than Aristotelian logic, this influence did not only come from ancient Greece. In fact, the period between the fourteenth and sixteenth centuries saw fresh translations of Islamic mathematical manuscripts into the Latin-speaking world. Both algebra and trigonometry were transmitted to medieval Christendom during this time. Biagio, for example, an Italian mathematician, living in the fourteenth century, presented simple algebraic operations as a part of practical arithmetic. In the fifteenth century, quadratic equations were solved by geometric means—as they had been in the Islamic civilization. These developments in Islamic science were transmitted primarily via universities in Italy. Studies in trigonometry were made from Arabic texts by Regiomontanus, a German astronomer who lived in Northern Italy in the fifteenth century.

Regiomontanus made explicit use of the sine function, stated the laws of the sines, and wrote a treatise on spherical trigonometry, from which Copernicus is supposed to have borrowed without acknowledgment. Regiomontanus built his own astronomical observatory, and is regarded as one of the first astronomers in Latin-speaking Europe who succeeded in treating astronomy as an exact science. The work of seventeenth-century scientists drew much from his astronomical work. In fact, Regiomontanus is known to have hinted at the notion of the motion of the Earth, long before Copernicus formulated the heliocentric theory. Regiomontanus stated, "It is necessary to alter the motion of the stars a little because of the motion of the earth."[30]

Communication of an experimental approach to testing theories, the importance of precision scientific instruments, and new developments in mathematics, beyond what was known in ancient Greece, began to reach Western Europe in the early part of the eleventh century. One of the first to recognize value in cultures outside Latin Catholicism was the Frenchman known as Gerbert of Aurillac. He went to a town called "Vich Catalonia," (the far northeast corner of the Spanish peninsular) to study Islamic mathematics and astronomy in the tenth century, a frontier town where there was good communication with the Muslims of Al-Andulas to the south. The library in the cathedral-school where Gerbert of Aurillac studied (under the care of Bishop Vich) and the one in the nearby monastery of Ripoll, were amongst the largest and best-equipped libraries in Europe at that time, benefiting from their proximity to libraries under Islamic rule. In Spain, Gerbert not only learnt about theoretical subjects such as the Indian decimal system and algebra, but he also came across astronomical instruments such as the astrolabe, and counting instruments such as the Chinese abacus, things that were unknown in Western Europe at the time. He also became aware of tables of accurate astronomical observations. All of this must have influenced him later when he went to study at the cathedral-school of Reims, where he distinguished himself in certain aspects of science. Historian Lynn Harry Nelson describes how Gerbert surprised his colleagues in Reims by carrying out experiments and by his use of superior mathematical methods. Nelson writes:

> He set himself to the task of building an organ with constant pressure supplied by water power. There had been organs before, but their air pressure had been generated by the organist pumping with his feet of an assistant pumping a large bellows. This one not only gave an extended steady level of sound, but its pipes were matched mathematically so that its harmonics were superior to anything heard in the West before. Gerbert had also mastered Arabic numerals and so could do calculations in his head that were extremely difficult for anyone thinking in terms of Roman numerals. He continued to study the abacus, and even constructed a giant one. He marked out the floor of the nave of the cathedral of Reims like an abacus and made a number of large disks to take the place of the abacus beads. He gathered some sixty-four members of the cathedral school to help him,

gave them sticks to push the disks, and sat in the organ loft from where he could see the entire floor. He would call out instructions, and his assistants would move the disks like a great game of shuffleboard. He was able in this way to deal with numbers both larger and smaller than had ever before been possible. He then wrote a book on the abacus that became standard in the new cathedral schools that were arising and revolutionized the study of mathematics in the West.[31]

As the above quotation demonstrates, Gerbert was immediately recognized by his contemporaries to be someone who took a new approach to learning, which involved both carrying out experiments and applying powerful mathematical methods to engineering problems. The important point to notice here, is that these aspects to his approach were not only missing in Latin-speaking Europe at the time, but they were also absent in the ancient Greek heritage. Both Aristotle and Plato did not prescribe the carrying out of experiments, nor did they have any knowledge of a decimal number system. The carrying out of experiments together with the use of more powerful mathematical principles was to later characterize the rise of modern science in the seventeenth century.

After his time at the cathedral-school of Reims, first as a student of logic, and later as its headmaster, Gerbert moved to Northern Italy as abbot of the monastery at Bobbio, then back to Reims as archbishop. Finally, he became Sylvester Pope II, in 999 CE. In all these different posts, Gerbet was, of course, able to contribute significantly to the study of science in Western European cathedral-schools. Historian David Knowles, writing on the influence of Gerbert of Aurillac, states, "The first great name in the history of medieval thought is that of Gerbert of Aurillac, who was master of the school of Rheims c.972, and ended his life as Pope Sylvester (999–1003 CE). In his writing on logic, and still more in his mathematical interests and his use of Arabian sources, he was the harbinger of many new things."[32]

Another medieval Western European important for publicizing a new approach to learning was the Englishman named Adelard of Bath. He was one of the first to introduce the Indian number system into the West, he translated the astronomical tables of al-Khwarizmi and the *Introduction to Astrology* of Abū Ma'shar. He wrote a short treatise on the *abacus* and astrolabe. He even disguised himself as a Muslim student in Cordoba, to obtain a copy of Euclid's famous book, "Elements," which he later translated and made available to the Latin-speaking world. He is also a significant figure in Western European history, because he credited the "Arabs" for teaching him not to blindly follow authority, but to use reason. Adelard of Bath declared:

Of course God rules the universe, but we may and should enquire into the natural world. The Arabs teach us that.[33]

In a conversation with his nephew, where Adelard of Bath explains what it is that he had learnt from the "Arabs," he states:

> We must first search after reason, and when it has been found, and not until then, authority if added to it, may be received. Authority by itself can inspire no confidence in the philosopher, nor ought it to be used for such a purpose.[34]

The main point here is that within the history of modern science, it was primarily the impact and transfer of Asian science and technology that caused medieval Latin-speaking Europe to emerge out from cultural obscurity, and that the influence of the philosophies of Plato and Aristotle were in comparison, at best, marginal. The whole preoccupation of late medieval Latin-speaking European scholars with the writings of Aristotle and Plato is better understood within the context of a debate in theology, in the effort to broaden exclusivist church doctrines, rather than being anything that has relevance to the rise of modern science.

CONCLUSION

This chapter has taken a multicultural perspective in summarizing the scientific and technological dialogue between Asian civilizations in the first millennium CE. It has given some examples to demonstrate that the contributions from Asia to the history of modern science are best understood within a multicultural framework, where cross-cultural exchanges between Indian, Chinese, and Islamic civilizations are taken into account. The chapter also examines the East-West dialogue in connection with the rise of modern science. It argues that, until relatively recently, most Western European accounts of the rise of modern science have tended to exaggerate the role played by the philosophies of Plato and Aristotle, while downplaying the influence of Asian civilizations. The chapter describes how religious-political rivalries in medieval Europe might explain why such a view arose and gained such widespread acceptance.

END NOTES

1. For China, see Joseph Needham's series, *Science and Civilization in China*, 1st ed. (Cambridge: Cambridge University Press, 1954); for Islamic civilization, see Seyyed Hossein Nasr, *Science and Civilization in Islam* (Cambridge, MA: Harvard University Press, 1968); and for India, D. M. Bose, S. N. Sen, and B. V. Subbarayappa, *A Concise History of Science in India* (New Delhi: Indian National Science Academy, 1971).
2. See Toby E. Huff's, *The Rise of Early Modern Science* (London: Cambridge University Press, 2003).

3. See Arun Bala, *The Dialogue of Civilizations in the Birth of Modern Science* (New York: Palgrave Macmillan, 2006).

4. Joseph Needham's massive achievement is embodied in the continuing Science and Civilization in China series, the successive parts of which have been published by Cambridge University Press since 1954. The home of the project is now the Needham Research Institute in Cambridge, which also houses a unique and growing library (the East Asian History of Science Library) used by scholars from East and West alike.

5. J. Needham, *The Grand Titration* (London: George Allen and Unwin Ltd, 1969), 62.

6. Ibid., 63.

7. A. Pacey, *Technology in World Civilization* (Cambridge, MA: The MIT Press, 1991), 16.

8. Ibid., 17.

9. J. Needham, *Clerks and Craftsman in China and the West* (Cambridge: Cambridge University Press, 1970), 131–132.

10. Pacey, *Technology in World Civilization*, 43.

11. D. Hill, *Islamic Science and Engineering* (Edinburgh: Edinburgh University Press), 229–231.

12. Needham, *The Grand Titration*, 79.

13. Needham, *Clerks and Craftsmen*, 16.

14. J. Ching, *Chinese Religions* (Maryknoll, NY: Orbis Books, 1993), 179.

15. J. Needham, *Clerks and Craftsmen*, 19.

16. Pacey, *Technology in World Civilization*, 16–17.

17. C. A. Ronan, *The Shorter Science and Civilisation in China*, vol. 1, An abridgement of Joseph Needham's original text, vols. 1 and 2 (Cambridge: Cambridge University Press, 1980), 38–39.

18. J. Needham, *Science and Civilization in China, Vol. 3, Mathematics and the Sciences of the Heavens and the Earth* (Taipei: Cave Books Ltd, 1986), 109.

19. Ibid., 43.

20. G. G. Joseph, *The Crest of the Peacock, Non-European roots of Mathematics* (London: Penguin Books, 1990), 347.

21. Needham, *Clerks and Craftsmen*, 28–29.

22. Pacey, *Technology in World Civilization*, 50–52.

23. Ibid., 46–49.

24. B. Russell, *History of Western. Philosophy*, counterpoint ed. (London: Routledge, 1984), 420.

25. A. Koestler, *The Sleepwalkers* (London: Pelican Books, 1968), 90–91.

26. D. C. Lindberg, *The Beginnings of Western Science* (Chicago, IL: University of Chicago Press, 2007), 160–161.

27. "It is natural to ask why Constantinople, where pure Greek culture had flourished without a break, and which would therefore seem at first sight the obvious centre for those interested in Greek Philosophy, did not become the Mecca of all Western scholars. The answer is complex and illustrates the larger question of the rift between East and West that existed all through the Middle Ages, that age-long and bitter misunderstanding between Greeks and Latins, which made close social and intellectual relationship impossible. To this must be added both the lack of a clear-cut programme

among Western scholars, who by no means agreed that they were searching principally for works of philosophy, and a lack of contemporary Greek interest in philosophy as a principal mental pursuit." D. Knowles, *The Evolution of Medieval Thought*, 2nd ed. (London: Longman, 1988), 168–169.

28. Koestler, *The Sleepwalkers*, 109.

29. Hill, *Islamic Science and Engineering*, 73–74.

30. Koestler, *The Sleepwalkers*, 212.

31. Lynn Harry Nelson, *Lectures in Medieval History 21: The Life of Gerbert Aurillac* (ca. 955–1003); D. J. Mabry, *The Historical Archive*, 1990–2009, http://historicaltextarchive.com/.

32. Knowles, *The Evolution of Medieval Thought*, 85–86.

33. Adelard of Bath, *Warburg Institute Surveys and Text XIV*, edited by C. Burnett (London: Warburg Institute, 1987), 16.

34. Adelard of Bath, *Dodi Ve-Nechdi*, (Natural Questions), translated and edited by H. Gollancz (Oxford: Oxford University Press, 1920), 98–99.

10

SCIENCE, TECHNOLOGY, AND
CIVILIZATION RECONSIDERED

Andrew Brennan

PLANTERS AND PIRATES

In his provocative book on the geography of thought (Nisbett 2003), Richard Nisbett distinguishes the centralized, settled, and agricultural peoples of ancient China, from the inhabitants of ancient Greece who busied themselves in fishing, hunting, and trading. Farming, he writes, "required people to cultivate the land in concert with one another," emphasizing the importance of cooperation and harmony. Large-scale irrigation required a network of social control, favoring an authoritarian and despotic polity of village elders, regional magistrates, and— ultimately—a central king or emperor (Nisbett 2003, 34). However, the early Greeks occupied an environment that favored "hunting, herding, fishing, and trade (and—let's be frank—piracy)." These activities require less large-scale cooperation, and when settled agriculture came to Greece (two millennia after it came to China), "many farmers were more nearly businessmen than peasants. The Greeks were therefore able to act on their own to a greater extent than were the Chinese. Not feeling it necessary to maintain harmony with their fellows at any cost, the Greeks were in the habit of arguing with one another in the marketplace and debating one another in the political assembly" (Nisbett 2003, 34–35).

It is helpful to have such an uncompromising statement of the conditions that, for Nisbett, produced ways of thinking in ancient Greece that in time gave rise to modern science in seventeenth-century Europe. Since the interdependence of the ruralist contrasts in Nisbett's story with the independence of the pirate, we can simply caricature the positions by referring to the two cultures as planters versus pirates. The planter mode of life—in Nisbett's view—is not one where we expect much by way of progress or science. In the European Dark Ages, when interdependence and the planter way of life characterized much of the continent, the results were pretty disastrous: "While Arab emirs discussed Plato and Aristotle and Chinese magistrates displayed their proficiency in all the arts, European nobles sat gnawing joints of beef in damp castles" (Nisbett 2003, 39).

Luckily, the Renaissance banished the horrors of the Dark Ages, and with it emerged a new sort of person, one with "curiosity and critical habits of mind." That person was the historical descendent of the Greek pirate, for "well before the fifteenth century, these values and the mentality that goes with them had been implanted in the European mind" (Nisbett 2003, 41).

The question about planters and pirates has been the subject of more subtle and nuanced studies than Nisbett's. Geoffrey Lloyd's work on authorities and adversaries, and—more recently—on ancient worlds and modern reflections, also starts with a picture of two cultures and two "mentalities" (see Lloyd 1990, 1996, 2002, 2004). Following the dubious authority of Alan Cromer, Nisbett contends that science as such was invented in Greece—and only in Greece (Nisbett 2003, 38, and compare Cromer 1993, 112–120). More cautious, Lloyd recognizes that none of the ancient traditions have science in anything like its modern guise, and instead they engaged in "science-like" activities, activities that indeed take a different shape in the different places. As Lloyd has repeatedly emphasized in his work, a major problem in approaching the description of what was happening in these societies that are so temporally, conceptually, and politically remote from us is finding the right words and ideas. It is not a simple matter to pose questions about these old cultures, for the forms of words and the concepts that we suppose in posing the questions are themselves likely to bias the answers we give. For example, there is nothing innocent in the question of what standard of truth was adopted by the ancient Greeks, or in ancient Chinese protoscience. Since it is hard even to start to answer these questions, it is with some qualms that we should approach the question of how to compare these standards (if they exist). Similar problems obtrude when we think, for example, about whether there were common beliefs across the ancient cultures, and whether these beliefs concerned a common world.

Lloyd and Nisbett offer two extreme approaches to dealing with problems in the history of ideas. While Lloyd is aware of the tradition in which Greek thought is adversarial where the Chinese are irenic, the former atomistic rather than holistic and committed to logic rather than to the tolerance of inconsistency, he still tries to understand these remote cultures in their own terms as far as that can be done. Extracting an idea from its surrounding cultural and social context risks anachronism, simplification, and a host of other distortions. Lloyd's conclusions are typically modest, qualified, and tentative. For example, in discussing

cross-cultural relativism and universalism with respect to the classifica-
tions of animals in ancient Greece and China, Lloyd concludes:

> The ancient classificatory endeavours we have been discussing bear many
> marks from the cultures that produced them, from their value systems and
> ideologies. But they also exhibit a certain plasticity, a certain openend-
> edness, which, in so far as it problematizes the notion of species itself,
> challenges a common assumption that underpins the debate between the
> cross-cultural universalist and the cultural relativist. Enquiry has always,
> to be sure, to be conducted with an explicit or a covert programme in
> view, and we have seen that the agenda of Greek and Chinese investigators
> differed in important respects. Yet the results obtained were not always
> predictable, nor always predicted by the ancient investigators themselves,
> thereby giving the lie to both extremist parties, both to those who pos-
> tulate a universal common sense underlying all zoological classifications,
> and to those who assume they are all determined straightforwardly by
> cultural factors. (Lloyd 2004, 117)

Nisbett, by contrast, relishes the recovery of notions that historians of
ideas have for some time considered to be discredited, such as "the Asian
mentality," "the Chinese mind," and indeed the simple opposition of
"Europeans" and "Asians."

Rogues' Gallery

Nisbett's findings on cognition are interesting, because they explain and
endorse some widely held stereotypes about "Western" and "Eastern"
modes of thought. Working not only with populations in East Asia,
Japan, and India, but with groups of US students of European and various Asian heritages, Nisbett and his collaborators have found a variety of
differences in cognitive processing in response to relatively simple tests.
These differences seem to show a significant difference between the cognitive capacities of Korean, Japanese, Indian, and Chinese people, on
the one side, and people of European descent on the other. These differences, in brief, are as follows (following the summary given in Brennan
2004):

1. *Continuity vs. discreteness*: the world as a collection of overlapping and
   interpenetrating stuffs (Chinese) vs. the world as a collection of discrete objects, categorized by reference to general properties (Greek);
2. *Field vs. object*: orientation toward the complexities of perceptual and
   conceptual fields as a whole (Chinese) vs. focus on a central focal
   object and its attributes (Greek);
3. *Relations and similarities vs. categories and rules*: focus on relations
   among objects, alteration by context (Chinese) vs. focus on categories
   that explain behavior independent of context (Greek);

4. *Dialectics versus foundational principles*: "The Chinese seem not to have been motivated to seek for first principles underlying their mathematical procedures or scientific assumptions...The Chinese did not develop any formal systems of logic or anything like an Aristotelian syllogism...In place of logic, the Chinese developed a dialectic...which involved reconciling, transcending or even accepting apparent contradiction." (Nisbett et al. 2001, 294)

5. *Experience-based knowledge vs. Abstract analysis.* "The Chinese... sought intuitive instantaneous understanding through direct perception" This resulted in a focus on particular instances and concrete cases in Chinese thought. By contrast, "many Greeks favoured the epistemology of logic and abstract principles, and many Greek philosophers, especially Plato and his followers, actually viewed concrete perception and direct experiential knowledge as unreliable and incomplete at best, and downright misleading at worst." (Nakamura 1964/1985, 171)

In terms of cognitive stance, there are basically two here—the *planter*, holistic, dialectical, and relational one ("Chinese") versus the *piratical*, analytic, formalistic, and object-oriented one ("Greek"). Nisbett repeatedly adverts to the frame and rod test, one in which a rod is viewed through a long box, square in sections (which from the viewer's point of view thus provides a frame around the rod). The experimenter can manipulate the frame separately from the rod, and the challenge for the experimental subject is to judge whether the rod is vertical independent of the orientation of the frame. "Asians," being more field- or environment-dependent in their judgments, have generally much greater difficulty than "Westerners" in assessing the position of the rod without bias from the frame. This kind of field-dependency is strikingly different between agricultural peoples and hunter-gatherers, with the latter having the same relatively low field-dependency as people in modern industrial societies (Nisbett 2003, 43). Not only does high field-dependency correlate with a liking for groups, a dislike of being alone and a tendency to sit close to other people while relaxing, it also explains why "Easterners" attend more to relationships, and less to objects, and "Easterners" see substances and change where "Westerners" see objects and stability (Nisbett 2003, 44).

Such findings on cognitive stance might seem a long way removed from studies of ancient Greece and China, and also a long way from any account of why modern science emerged in the way it did in the seventeenth century. But Nisbett, and others who have been impressed by these findings in cultural psychology, think there is a relatively clear argument that can be used to link the findings on cognition with questions about the birth of science. A group of philosophers and cognitive scientists have drawn attention to how Alison Gopnik's theory of "the scientist in the crib" might explain the rise of modern mathematically based sciences (see Faucher et al. 2002; Gopnik 1996). In Gopnik's view, the wealth of information available to thinkers in Europe due to improved communications and openness during the Renaissance stimulated the universal and innate theory-generating mechanisms that humans are born with. For Faucher et al., Nisbett's work is an important corrective to Gopnik's innatism and her lack of attention to the way that culture profoundly shapes cognition. Along with an appeal to the emergence of important norms of evidence in the seventeenth-century Europe—norms that themselves were spread by cultural means—Faucher and his colleagues attribute a major role to the culturally inculcated cognitive mechanisms described by Nisbett as providing a significant part of the explanation for why early modern Europe was the focus of the new sciences. After all, they argue, Chinese thinkers also had a wealth of new information available to them at the time of the European Renaissance, and they too had good communications. But modern science emerged not in China but in Europe.

The appeals on behalf of psychology seem to merit a place for it in the rogues gallery of theories that already exist and claim to account for why the mathematical sciences emerged when and where they did. As I have pointed out in other work, it is hard to take Nisbett's so-called geographical claims seriously, for his labels "Asian" and "Western" do multiple jobs (Brennan 2004). First, they can identify modes of attention, thought, and perception that fit the dual pattern just given—what I've called "planter and pirate cognitive styles." But on the other hand, they also identify national, racial, or "geographic" stereotypes, stereotypes that the reported research itself undermines. Not all Asian peoples score consistently on the planter side of the divide, and not all Europeans count as pirates. In some of the studies that Nisbett and his colleagues cite, the planter cognitive feature is present only in 60 percent of Asian respondents. Put another way, there are 40 percent pirates in these samples. That is, in Nisbett's terms, 40 percent of these Asians are actually "Westerners." To add further confusion, Nisbett himself points out that sometimes the Germans, French, and Italians are less piratical in orientation as compared to other "Westerners." In analyzing a study discussing responses to a question seeking to identify whether a company is viewed as a kind of organism or rather as a system of separate components, Nisbett writes:

> About 75 per cent of American chose the first definition, more than 50 per cent of Canadians, Australians, British, Dutch and Swedes chose that definition, and about a third of Japanese and Singaporese chose it. Germans, French and Italians as a group were intermediate between the Asians and the people of British and Northern European culture. Thus

for the Westerners, especially the Americans and the other people of primarily northern European culture, a company is an atomistic modular place where people perform their distinctive functions. For the Easterners, and to a lesser extent the eastern and southern Europeans, a company is an organism where the social relations are an integral part of what holds things together. (Nisbett 2003, 84)

Notice the looseness of expression here. Neither France nor Italy is in eastern Europe, and East Germany no longer exists. So Nisbett has either made a slip of the pen or a gross geographical mistake. Now consider the sentence beginning with the word "thus." More carefully expressed, it might have read as follows: "For some Westerners, and for many north Americans and other people of primarily northern European culture...etc etc." But even this version would be misleading. For it is only a slim majority (just more than 50 percent) of the north Europeans who hold the atomistic conception of the company. It would be equally correct to say that for many people of primarily northern European culture (more than 40 percent), a company is an organism where the social relations are an integral part of what holds things together. Moreover, Galileo—an early modern scientist with whom Nisbett associates analytic and logical thought—was born in the Duchy of Florence (located in what is now called Italy). His piratical, logical, and analytic cognitive style seems to run against the trend of the Mediterranean peoples whom Nisbett locates as midway between being "Asians" and "Westerners." Of course, maybe the sixteenth- and seventeenth-century Florence was more "Western" than the Italy of today. But perhaps it would be best to stick to the *pirate* and *planter* labels to avoid the risk of muddles and of thinking that France and Italy are both a bit Western and a bit non-Western at the same time.

Taken in this way, Nisbett's claim seems to be that the birth of modern science was abetted by piratical rather than planter thought. In one way, this is almost a trivial claim: for modern science can simply be characterized as the breakthrough that resulted from applying analytical, logical, and reductive thinking to problems in astronomy, mechanics, and optics. Moreover, much of the basic work in mathematics and astronomy on which Copernicus, Newton, and Galileo drew was carried out in the great tradition of Arabic thought (a tradition on which Nisbett and his colleagues have apparently not carried out much study). Even if not "Western" in any cultural or geographic sense, the Arabic tradition can be thought of as piratical to the extent that it involved analytic cognitive strategies. Likewise, given the importance of mathematics to the newly emerging sciences, the significance of the algebraizing of geometry and the importance of the Indian number system (Bala 2006, 166–174), there must have been pirates in the Indian traditions too.

But since no cultural tradition is 100 percent piratical or planter, it is not clear just what work is being done by the appeal to cognitive styles as explanatory factors in debates about the birth of modern science. Modern science, as everyone agrees, did not emerge in China. Yet there are famous pirates to be found in the Chinese intellectual tradition too—most notably Mozu and his followers of the fifth and fourth centuries BCE, and the eleventh-century experimentalist Shen Kuo. As Nathan Sivin, among others, has pointed out, Shen Kuo's treatment of astronomical and other matters is surprisingly systematic and modern, though no consistent thread of what we would now call "scientific thinking" runs through his works (Sivin 1988). Instead, systematic and insightful remarks sit side by side with commentary on political rivalries and court intrigues. So Shen Kuo is part pirate and part planter. In fact, as Nisbett's own research shows, every sample from every culture contains both pirates and planters. Even if there is a planter mainstream in a given culture, there is no reason to believe that its best thinkers will match the mainstream in their cognitive styles. And indeed, there is no evidence for supposing that the most cognitive styles within any culture are also typified by its best poets, painters, philosophers, or scientists.

IDEOLOGY AND MATERIALISM

Nisbett's use of cultural psychology is meant to suggest that patterns of thought, perception, and cognition that are culturally inculcated can make it harder or easier for thinkers and groups of thinkers to develop systematic science. For him, the mentality of the Chinese was not such fertile ground for theory development. By contrast, we are meant to think that the mentality of people in Italy, France, and other parts of Europe during the development of modern science was well fitted for the task. Suppose then that piratical cognitive influences held sway in the minds of Copernicus, Galileo, Descartes, Newton, and the other notable figures in early modern science. How much weight should be given to the cognitive against all the other kinds of factors that can also be adduced in explanation of the birth of science? One great schism that separates out accounts of the great event is between those explanatory stories that are *ideational* one way or another, and those that are *material*. Ideational accounts focus on the economy of ideas in circulation at a time and accessible from earlier times, looking for ways in which certain ideas, or forms of thinking, can have inspired others. The idea of the atom, and atomic theorizing in general, is an example of the ideational perspective. Scholars can trace the rise and fall of atomism (or atomisms) through times and cultures, looking for the seeds of new ideas, ways of blending ideas about atoms with other protoscientific ideas.

By contrast, material accounts of the rise of science will focus on features of the environment against which various sets of ideas were explored, investigated, developed, or discarded. Perhaps, one of the most plausible material accounts of the rise of scientific thought is the suggestion that at times of conflict, disturbance, and warfare, new constellations of thinkers and ideas appear, people meet who would not otherwise have interacted, ideas and theories come under challenge, and innovation and creativity is thereby stimulated. In the history of Chinese thought and innovation, for example, it is striking how the spring and autumn and warring states period provoked stirrings in medicine, technology, and philosophy, as was also the case in the Song dynasty, where the flowering of arts and technology followed hard on the heels of the unsettled five dynasties and ten kingdoms period. But many such accounts are overdetermined by a plethora of material causes, no one of which can be singled out for special treatment. Warfare and political upheaval, for example, often involve movement and displacement of populations: but so do the impacts of other changes in transport and technology, not to mention population growth itself that goes in stops and starts.

The cultural psychologists span both the material and ideational schools of history. Like the *Annales* historian Fernand Braudel who claimed that the "'mentality' of France was contained within its physical, social and economic history" (see Oswyn Murray's introduction to Braudel 2001), Nisbett sees cognitive processes as nested within a large field of social, economic, and ultimately environmental situations. We have already noted his belief that settled farming communities produce less by way of cognitive dissonance than hunting and trading ones, where there is a greater impetus to the "development of cognitive procedures, including formal logic, to deal with the dissonance" (Nisbett 2003, 31). Oddly enough, Nisbett does not mention the possibility of civil strife, warfare, sudden environmental changes, population displacement, and other material variation driving societies and cultures toward the development of new cognitive procedures, even though such possibilities are compatible with his approach.

In general, the ideational and the material seem to be complementary to each other. If new cultural norms, and new cognitive procedures, emerge at certain times, these may well be in response to social, political, and environmental upheaval. But without an existing economy of ideas to draw upon, and without a range of existing techniques, ideas, and theories, there would be no food to sustain the next step. As Jacques Barzun noted, "All men, including the genius, live by others' ideas" (Barzun 1959, 259). A history of science that dwells only on the material circumstances of its production is thus just as incomplete as an ideational approach that neglects the processes by which certain ideas come to the fore at one time rather than another. Aiming for completeness here may be to aim at the impossible. Perhaps we serve ourselves best by looking for helpful—but necessarily incomplete—accounts whether ideational, or material or perhaps selectively combining elements of the two approaches in interesting new ways. To accept incompleteness is not to acquiesce in failure. After all, there is likely no single question posed by the following words: *why did modern science emerge in seventeenth century Europe?* Instead, there are many tangled questions, some ideational, some material, some—like the political and economic ones—sharing in both.

Arun Bala's book provides a fascinating guide to the ideational side of the many questions we face (Bala 2006). In his conclusion he writes that "Greek science did not really play the unique role now assigned by historians to it. It came to be seen as the only predecessor of modern science simply because it possessed a sufficiently rich reservoir of ideas for historians to construct it as such by selective appropriation of themes guided by the hindsight provided by modern science" (Bala 2006, 179). Read in one way, this claim is compatible with Georges Canguilhem's notion that the history of certain concepts can only be written once they have been developed into a mature form within a science. Finding reservoirs of ideas, as Bala points out, will not be enough to give even an epistemological or philosophical "history" of modern science. Think of paradigms—in a sense close to one of those used by Kuhn—as something that unites communities of investigators by supplying common models, exemplars and the like (Kuhn 1970). Such paradigms are at least half-recognized by the researchers who use them, and their persistence can be encouraged by institutional arrangements.[1]

At least for some times and in some places, the work of earlier thinkers has provided paradigms for research and such paradigms have thus bound disparate thinkers together, however loosely. In his work on the history of medicine and the other sciences, Canguilhem argues that before the emergence of a mature science, there exist one, or more, scientific ideologies. These are sets of beliefs and practices that hold a place, so to speak, for occupation by the mature science. Yet, once the science emerges it nearly always turns out that the place it occupies fails to coincide with the territory formerly marked out by the scientific ideology. Thus, for example, until electricity was subject to thorough scientific regimentation, a whole host of effects and theories were associated with electrical phenomena. Once a mature science was in place, however, some of the phenomena previously described as "electrical" were no longer so regarded, and things not previously associated with electricity were now revealed as electrical phenomena. In the dynamics of emerging science, this is what Canguilhem, following Bachelard, calls an *epistemological break* between the scientific ideology on the one hand and the science on the other (Canguilhem 1988). Another way of thinking about these scientific ideologies would be to regard them as prescientific paradigms.

Put in this alternative terminology, Canguilhem's claim is that the prescientific paradigm typically marks out a different territory from that occupied by the later science. The progression of science is then marked by a succession of paradigms, at least some of which will involve restructuring and reinterpreting material described in terms of earlier paradigms.

From this point of view an ideational history of science can be constructed by tracing the history of a number of earlier scientific ideologies, each one itself systematic and theorized to a certain degree, but also at the same time making claims beyond its current capacities (Canguilhem 1988). What Bala calls the "dialogue of civilizations" thus takes its place as a contribution to the ideational story, one that involves much more than indicating the presence of existing reservoirs of ideas. Moreover, as historical epistemologists have implied, there is a distinction between what we might regard as a philosophical or conceptual history of science and other kinds of history. I will just mention the philosophical version of history to be able to lay it aside in the conclusion of this chapter. Conceptually, modern mathematical science reveals the residues of ideas developed in China, Greece, India, the Arab lands, and many other places. Using contemporary science as a starting point, we can try to appraise the achievements of earlier people and groups. The task of the sciences is not yet finished. There is always more to learn, new paradigms to develop, corrections to be made, and dogmas to be overthrown. But taking the present state of the sciences as giving us a goal toward which we can imagine earlier thinkers aiming, we can ask how well they did, how far they got. With our newfound knowledge, we can express amazement at Greek geometry and its applications, wonder at the technical prowess of the ancient Chinese, marvel at the Indian number system, and be stunned by the inventiveness of the Arab astronomers in overcoming the limitations of the Ptolemaic model. Those who approach the history of science not so much as historians, but rather as scientists, philosophers, and others interested in concepts, systems of ideas, and structures of implication, will find much to marvel at and enjoy. The reason such a study is conceptual or philosophical is that in viewing past achievements through the lens of the present, it focuses on the economy of ideas and concepts as it has been regimented over time. Notions such as "influence" or "cause" are not needed in such a study, for it is concerned more with logical or conceptual relations than with questions of who spoke to whom, what development influenced what other developments, and so on. Many comparative studies of ideas are of this sort. Comparing classical Greek and Chinese systems of measurement, or looking at the philosophy of place implicit in traditional Chinese and Ayurvedic systems of medicine (see Brennan 2002) are examples of this kind of work. The results of such work are often illuminating, using the remote tradition as a mirror to our own, or using two remote traditions as casting conceptual light on each other. In laying aside such work in the present chapter, I am not suggesting that it is unimportant or dangerous. What is important, though, is to distinguish this kind of investigation of the epistemology and conceptual economy of earlier scientific ideologies from the discussion of what influenced what, and what caused what.

By contrast, for the historian who is less concerned with investigating and comparing earlier conceptual economies, changes in systematic studies of nature in what we now think of as Europe occurred at different times through a complex set of contingencies, influences, and accidents. Studies waxed and waned depending on political conditions and on the work of very different individuals, people who often had little idea of what they were doing, and where they were going, and certainly—if we think of Newton, Kepler, and Copernicus—were as much engaged in magical, religious, and correlative modes of thinking as in logical, mathematical, and causal thinking. Indeed, just what sort of thinking the early modern scientists were engaged in is not an easy question to answer. Changes in thought resulted from appropriating inspirations from many sources, whether from the Greeks, the Sephardic Jews, Chinese sages, or Islamic mathematicians and astronomers. Even without considering the material conditions that existed, the ideational trajectories that can be historically tracked, classified, and plotted are stunning in their complexity. Now add to all this the economic conditions, agricultural productivity, aggressive rivalries, and warfare that also have a part to play in explaining not only the changes that occurred but also the time-lag between the proposing of an idea and its appearance as a part of systematic knowledge of the world. The jumble of material and ideational, the openings created by the great periods of instability and warfare, the demand for advances in military technology—these all combine to produce the contingencies that made up the so-called scientific revolution.

HISTORICAL EPISTEMOLOGIES

This chapter started with cultural psychology, as a provocation. For thought about the history of science and history of ideas in the twentieth century has often adopted quite the opposite tack from the one taken by Nisbett. For him, logic, analytic thought, and the other aspects of piratical cognition were present in ancient Greece and rampant in early modern science. The cognitive processes that I have labeled "piratical" are the one whose emergence was forced, he thinks, by encounters with cognitive dissonance, and which in turn gave rise to science as we know it. This account stands in stark contrast to a theme in twentieth-century history of science exemplified by Gaston Bachelard and developed by Canguilhem, the notion of "applied rationalism" (Bachelard 1951; Canguilhem 1988). For Bachelard, logic and rationality are not so much the *sine qua non* of science, but rather the results of it. There is not one rationality, but rather a series of regional rationalisms, each associated with the development and progress of a branch of the sciences. Inspired by the developments in physics of the first half of the twentieth century, Bachelard regarded contemporary physics as having broken with earlier physics and as having thereby forged new forms of rationality.

There is a suggestion here that can be applied in assessment of the cultural psychologist's position as put forward by Nisbett. For him, cognition is not ahistoric, not some kind of analytic rationality given independent of the situation of the thinker. In fact, Nisbett regards logical thought as a product of cognitive processes that are in turn the product of historical and environmental circumstances. He believes that piratical cognitive processes were present in ancient Greece, at scattered times during the Renaissance and came to the fore during the last three centuries. But if the suggestion inspired by Bachelard and Canguilhem is right, it is highly unlikely that the very same cognitive processes that characterize modern educated people were present in the seventeenth century, let alone two and a half thousand years ago. The scientific ideologies of the Greeks, like those of the Chinese, the early Arab astronomers, the Indian mathematicians—these, and other attempts at understanding the world, were stages in the historical emergence of the scientific rationality we now have. The analytic, reductive, and atomistic cognitive style that shows up in tests administered by Nisbett and his colleagues is the product, not the cause, of modern science. According to this account, Nisbett has got things quite the wrong way round. Instead of the pirate's cognitive process driving the emergence of science, they are the product of science. By reading these processes back into the minds of ancient Greeks, and taking them as indicating characteristics of the culture as a whole, Nisbett is guilty of an anachronistic folly, to say the least.

From the view point of historical epistemology, the early theorizing in all the ancient traditions constituted stages in the emergence of the contemporary sciences. The study of these protoscientific and prescientific investigations is often helpful as a mirror to our current research and speculations. As they developed into paradigms and scientific ideologies, these investigations helped lay out a trajectory, one not seen as such by those involved in the earlier investigations. For it is in the nature of pioneers of science to make claims that will ultimately be shown false, to map a territory for their studies that will sooner or later have its borders redrawn. Scientific ideologies remain, however, fascinating objects of study, and contain material that—in quite unpredictable ways—can stimulate new ideas and perhaps new breakthroughs. While Nisbett sees the cognitive structures of settled rural people as posing an impediment to the development of science—an impediment he thinks still exists nowadays—the historically cautious theorist will wonder if the limits of analytic and rationalistic theorizing will one day be transcended by more holistically inspired forms of science, thus constituting an epistemological break every bit as exciting as the trauma in physics induced by the advent of relativity theory and quantum mechanics. However, and equally cautiously, it may be that we are in the infancy of forms of science that we wrongly regard as already mature, and that what we take for a developed science is itself one more scientific ideology, making claims beyond its capacities and misidentifying its core subject-matter.

There is an unfortunate tone to the question: Why did modern science not emerge in China, or in India, or...? Often we ask "why...not...?" questions in situations where we are being critical of an omission. Why did I not see the proof? Why did you not clean your teeth? Why did I not stop her setting out in the storm? All these questions have a normative implication—that I ought to have seen the proof, you ought to have cleaned your teeth, I should have stopped her going out in the storm. So the implication of "Why did modern science not emerge in country X?" can seem to be that modern science should have emerged in that place. But there are no "shoulds" in the history of science and creativity. The self-congratulatory tone of writings on the history of ideas by European-based authors reflects their own (perhaps unconscious) prejudices, just as Nisbett's reference to the ancient Greek agriculturalists as being "more nearly businessmen than peasants" hints at an ideological orientation for his work of which he himself is likely unaware. Ideologies and prejudice aside, my suggestion is that the "why...not...?" questions make no sense. About the related positive question, I previously wrote: "the question "Why did science as we know it emerge in Europe, not in China?" is empty if construed as asking why some geopolitically-defined population has been the host of an analytic culture which carried a nascent science to its full maturity in modern mathematical form" (Brennan 2004, 219), and I see no reason to change this view now.

Political contingencies, population movements, and warfare would seem to carry more weight in explaining why certain ruptures in the fabric of knowledge occurred at one place rather than another than do speculations about the cognitive structures of individual thinkers, or of cultural traditions to which they belong. To note this point is not to elide the importance of the distinctions between interdependence and autonomy, holism and reductionism, attention to relations rather than principles, appeal to experience rather than abstracta—all these can be ingredients of worldviews that people bring to bear on issues in science, politics, management, ethics, and life in general. Cultures within which certain worldviews flourish and are practiced can inculcate a kind of *habitus* (Bourdieu and Passeron 1977) that may well show up in the cognitive psychology test. Becoming aware of this may itself stimulate new ideas, open up new forms of study, especially in medicine, ecology, and parts of the biological sciences. To this extent, the study of cultural psychology may itself prove useful as a stimulus to new ways of thinking.

Cultures themselves are, and let us hope always will be, places of tensions, disputes, and fashions. Cultures nowadays are all likely to be affected by the success of the sciences themselves, and the ways of thinking and perceiving that we now consider to be most "natural," will be in part a reflection of the modes of thought and cognition that are products of the contemporary sciences. If there is anything at all to be learned from the history of the sciences, and that can safely be projected onto the future, it is that some places and times may allow for more openness, debate, and flexibility than others. The European Renaissance and Reformation was just such a time of opening and opportunity. Copernicus drew heavily on Islamic astronomy and mathematics, and in many other ways the Renaissance and the seventeenth century allowed the genius found in a variety of traditions—including Chinese, Indian, and Islamic ones— to interact, cross-pollinate, and flower. We can, perhaps, look forward with a mixture of apprehension and expectation to future openings that in turn provoke rupture in the cultural fabric, leading to experimentation and conflict involving diverse values and cognitive norms. Such times are not to be reduced to simple formulas and single-factor explanations. As often remarked by historians of science, key figures have the rather unnerving characteristic of looking in different directions at the same time. Sivin, as already noted, comments on Shen Kuo's inability to maintain a clear scientific thread through his writings. For a more recent example, think of Paracelsus who rejected humoral pathology in favor of chemical remedies for disease (Magner 1992, 171), yet was also willing to rely on supernatural healing. We do not yet know how consistent or inconsistent our own age will seem to future historians of ideas, but perhaps the most important thing that we can learn from the past is to be modest in our claims about what we have learned so far, and what we are still to learn.

END NOTE

1. As an aside, note that universities can also play a part in the material history of science. For they are not just reservoirs of ideas, but can be places within which paradigms can be forged, challenged, and passed on from teachers to disciples. The great medieval universities may thus have a part to play in one of the stories about how modern science emerged—and this despite the fact that academies and universities have often been places dedicated to the conservation of traditional lore or esoteric doctrine rather than centers of critique, innovation, and invention.

Bibliography

Bachelard, G. 1951. *L'activité rationaliste de la physique contemporaine.* Paris: Presses Universitaires de France.

Bala, Arun. 2006. *The Dialogue of Civilizations in the Birth of Modern Science.* New York: Palgrave Macmillan.

Barzun, Jaques. 1959. *The House of Intellect.* New York: Harper.

Bourdieu, P., and J. C. Passeron. 1977. *Reproduction in Education, Society and Culture.* Translated by R. Nice. London: Sage.

Braudel, F. 2001. *Memory and the Mediterranean.* Translated by S. Reynolds and edited by R. de Ayala and P. Braudel. New York: A. A. Knopf.

Brennan, Andrew. 2004. "The Birth of Modern Science: Culture, Mentalities and Scientific Innovation." *Studies in History and Philosophy of Science* 35: 199–225.

———. 2002. "Asian Traditions of Knowledge: The Disputed Questions of Science, Nature and Ecology." *Studies in History and Philosophy of Biological and Biomedical Sciences* 33: 567–81.

Canguilhem, Georges. 1988. *Ideology and Rationality in the History of the Life Sciences.* Cambridge, MA: MIT Press.

Cromer, Alan. 1993. *Uncommon Sense: The Heretical Nature of Science.* New York: Oxford University Press.

Faucher, L., R. Mallon, D. Nazer, S. Nichols, A. Ruby, S. Stich, and J. Weinberg. 2002. "The Baby in the Lab-Coat: Why Child Development Is An Inadequate Model for Understanding the Development of Science." In *The Cognitive Basis of Science,* edited by P. Carruthers, S. Stich, and M. Siegal, pp. 335–362. Cambridge: Cambridge University Press.

Gopnik, A. 1996. "The Scientist as Child." *Philosophy of Science* 63, no. 4: 485–514.

Kuhn, T. H. 1970. *The Structure of Scientific Revolutions.* 2nd ed. Chicago, IL: Chicago University Press.

Lloyd, G. E.R. 2004. *Ancient Worlds, Modern Reflections: Philosophical Perspectives on Greek and Chinese Science and Culture.* Oxford: Clarendon Press.

———. 1996. *Adversaries and Authorities.* Cambridge: Cambridge University Press.

———. 1990. *Demystifying Mentalities.* Cambridge: Cambridge University Press.

———, and Nathan Sivin. 2002. *The Way and the Word.* New Haven, CT: Yale University Press.

Magner, L. N. 1992. *A History of Medicine.* New York: Marcel Decker.

Nakamura, H. 1964/1985. *Ways of Thinking of Eastern Peoples.* Honolulu: University of Hawaii Press.

Nisbett, Richard E. 2003. *The Geography of Thought: How Asians and Westeners Think Differently—and Why.* London: Nicholas Brealey.

———, K. Peng, I. Choi, and A. Norenzayan. 2001. "Culture and Systems of Thought: Holistic Vs. Analytic Cognition." *Psychological Review* 108: 291–310.

Sivin, Nathan. 1988. "Science and Medicine in Imperial China—The State of the Field." *The Journal of Asian Studies* 47: 41–90.

III

Forging New Knowledge

11

TOWARD CONSTRUCTING POST-POSTMODERN TWENTY-FIRST CENTURY SCIENCES: THE RELEVANCE OF CLASSICAL CHINESE MEDICINE

Keekok Lee

INTRODUCTION

The world is on the cusp of great changes—political, economic, and physical. Economic and political power is shifting from the mature economies and democracies to rapidly developing countries such as China and India. Such fundamental shifts may have implications for an epistemological shift from "West" to "East." The philosophical paradigm behind the modern revolution in science since the seventeenth century underpins the rise of Western supremacy for the last two centuries or so. That paradigm, its entailed methodology, including its understanding of causal agency and logic are being severely challenged by the crisis of global climate change and sustainability, as well as more spectacularly the meltdown of the global financial system. George Soros[1] has correctly argued that the cause of the latter lies in the inherently flawed nature of the causal paradigm behind market fundamentalism; an analogous flaw appears to have retarded the recognition that a climate change crisis confronts humankind. Soros proposes the alternative paradigm of reflexivity, which this author sees but as a subspecies of a more general paradigm involving a "post-postmodern concept of cause and logic" that is non-Humean and non-Aristotelian, being nonlinear, necessarily multifactorial and reciprocal, nonreductionist in character, and holistic in its metaphysics. One may argue that this kind of "ecosystemic science" was pioneered, though not perhaps self-consciously, by the ancient Chinese in their theory and practice of "Traditional Chinese Medicine," although for the purpose of this chapter, the term "Classical Chinese Medicine" is preferred. This chapter argues that the post-postmodern[2] twenty-first-century world would, increasingly, in its various fields of intellectual endeavor, have to attend to this very ancient, yet contemporary understanding of causality and logic with which it is entwined, should it wish to solve problems in economics and finance, global warming, and sustainability or, indeed, in medicine itself.

CONTRASTING CAUSAL PARADIGMS: OLD AND NEW

Let us begin by looking at the new ecosystems paradigm and contrast it with the old, Humean linear paradigm to which it hopes to add, though not to supersede. The Humean paradigm involves a notion of cause that can be seen as assuming the following:

> Whenever (event) A occurs (event) B occurs; A precedes B in time (or occurs simultaneously with B); A and B are spatially contiguous; there is no necessity in nature, just the plain observation that A and B are constantly conjoined.

This is also referred to as the uniformity of sequence analysis of cause, or the billiard ball account. Such an account assumes that the causal arrow points in one direction only, from A to B, B to C.

By contrast, the ecosystems rival paradigm is non-Humean and non-linear and emerges more prominently in the twenty-first century. It involves a "reciprocal" notion of cause as being multifactorial with more than two interacting agents, involving both negative as well as positive feedback mechanisms so that the system, when it reaches equilibrium, is not a mere return to the old under negative feedback, but reaches a new and different level of equilibrium. Also causal changes, being embedded within a system that is dynamic, take place continuously, and when certain conditions obtain, the system switches nonlinearly, that is, suddenly and without warning, to a different and new equilibrium that itself is necessarily temporary, given that the system is a dynamic one.

As this ecosystem paradigm is much less familiar, we must elaborate a little. The reciprocal notion of cause implies that when an agent A acts on agent B, producing an effect on B, B in turn can produce an effect on A. A quick illustration: A (either a human or an animal) finds itself itching in a particular part of the body; to ease the itch (B), A scratches B; A's scratching, far from easing the itch, increases it as scratching makes the skin react with greater ferocity than in the absence of such intervention. When the itch intensifies, A resorts to scratching it even harder, and so on, until the skin becomes so raw and so bloody that A can no longer scratch B.

The term "ecosystematicity," though infelicitous, may, nevertheless, be apt, as ecological studies of ecosystems, in the past few decades, have amply demonstrated the relevance of the concept of "causal reciprocity." Any ecosystem in real life is shaped by numerous agents.[3] These are often too many to be exhaustively identified as individuals, so for the sake of simplicity, we can resort to enumerating them in terms of classes or groups. Three main groups may be identified: living agents that are human, living agents that are nonhuman, and nonliving agents.[4] Theorists disagree about what shaped the composition of ecosystems on our planet. Most hold that without exception they involve all three types of causal agents; others, a minority,[5] hold that although it is getting more difficult to find ecosystems that can be said to be free from the influence of human intervention, it nevertheless makes sense to talk about ecosystems in the absence of human presence and manipulation. However, most theorists agree that all ecosystems on Earth include nonhuman living as well as nonliving agents. In a study of any ecosystem, the crux of the matter lies in the causal relationships that occur between the biotic, in whatever form, and the abiotic. This holds true at any level of investigation, whether the ecosystem is micro or macro—a handful of soil is as much an ecosystem as a virgin forest that may be the size of England and Wales combined.

The causal relationships between the interacting agents involved are reciprocal. Take the following example: a hairline crack exists in a rock (A); water (B) enters the crack, turning into ice, thereby enlarging the crack in the process; a seed floats by, lodges itself in the crack and grows (C); C together with B cause A to widen, which in turn permits more water and ice to enter, giving more space for C to grow by widening the crack still further and thereby allowing more rain and frost (B) to enter and erode it, and so on.

Global climate change[6] involves many, infinitely more numerous causal factors than the example cited above. The abiotic minimally includes Earth's atmosphere, carbon dioxide and other greenhouse gases, oceans, particulates and aerosols in the atmosphere, clouds, and water vapor. The biotic minimally includes humans who release carbon dioxide through burning fossil fuels; cattle and humans releasing carbon dioxide when they breathe, and methane when they fart and defecate, and when humans grow paddy rice; young trees absorbing carbon dioxide from the atmosphere, and decaying or felled trees releasing the gas. Large-scale deforestation without adequate replanting upsets the balance between the uptake of carbon dioxide and its release.[7] To grasp the phenomenon of global warming from the causal point of view, it is best to regard Earth and its atmosphere as a single ecosystem in which all the relevant agents are interacting in a reciprocal manner. An illustration of this kind of causality at work is "run-a-way greenhouse gases": excess carbon dioxide in the atmosphere (A) causes atmospheric temperature to rise (B); warmer air is able to retain more water vapor than cooler air (C), thereby causing atmospheric temperature (B) to increase even more, which in turn causes greater retention of water vapor (C), reinforcing the greenhouse effect of the excess anthropogenic carbon dioxide in the atmosphere.

Apart from reciprocal causality, another difference between the ecosystem and Humean competing causal paradigms is the notion of nonlinearity. To illustrate this we can look at the proverbial straw that broke the camel's back.[8] Assume that a straw weighs 1 g and that the camel can normally bear a weight of 50 kg. In other words, the camel can carry a bale of 50,000 straws. Up to the loading of the 50,000th straw, the animal stands firm; yet it collapses when the 50,001th straw is added. On the Humean linear model, such a phenomenon is neither foreseeable in principle nor intelligible, since it presumes that small causal changes will also produce small changes in the effects generated. On the nonlinear, ecosystems model, it can be both foreseen and explained. This is because, according to the linear model, the causal impact of each straw has to be regarded in isolation from the causal impact of the other straws piled on the camel's back. Each straw in isolation would be expected to produce no untoward effect on the camel. Yet the 50,001th straw, which weighs exactly the same as each of the others preceding it, has the dramatic effect of making the camel collapse. This cannot be explained by a linear understanding that regards causes and effects to be merely additive or subtractive, ignoring cumulative causes and effects. Such a linear orientation fails to recognize that the causal impact of the 50,001th straw on the camel's back is not the same as the causal impact of the 1st, 2nd, or 50,000th straw. The total causal impact on the camel of 50,000 straws is not simply the sum of the causal impact of each preceding straw taken separately and in isolation from other straws. This is what makes the camel collapse with the addition of the next straw.

Indeed, on the nonlinear model, the camel could collapse even well before the 50,000th straw was loaded on to its back if we understand cause not merely in terms of an event but also in terms of standing conditions, which are part of the systemic boundaries of the phenomenon under study. In such a case, what is causally relevant is not simply the fact that straws are being piled onto the camel's back, but also that these straws are being piled onto the back of a particular camel with its own specific condition of health, age, and state of exhaustion, which may be said to constitute the standing conditions. Therefore, how much a camel can bear depends on its general physiological conditions at the time it enters the causal equation.

It follows from the above that the camel-straws relationship must be regarded as a system that is dynamic, rather than static. Although factors such as exhaustion, hunger, disease, and old age would not on their own distress the camel, yet acting together synergistically, the distress could be magnified several times over, so that even before the 50,000th straw was piled on to its back it could collapse. In contrast, to the linear account, where the impact of each straw is constant and unchanging irrespective of the state of the camel, the nonlinear alternative is historical and dynamic. The latter regards the straw and its causal impact to be part of the standing conditions and systemic boundaries that produce causal effects. What constitutes the effect at time t_1 can be part of the cause at time t_2. This explains why the state of the camel bearing 50,000 straws at t_1 together with the 50,001th straw causally brings about the collapse of the camel at t_2.[9]

CLASSICAL CHINESE MEDICINE

This has relevance to understanding the seemingly outrageous claim on the part of this chapter that classical Chinese medicine (CCM) could remotely be called a science. This is because CCM embodies a non-Humean causal and logical paradigm. Such an understanding of CCM also points to an answer to the so-called Joseph Needham Question: "Why did the West overtake China in science and technology when China led the world up to the beginning of the 17th century?" Or "Why did Western Europe develop modern science in the 17th century, but not China?" For nearly five decades scholars have been stirred to answer this question, including Needham himself, albeit quite halfheartedly. Scholars agree that though many answers have been proffered, none has proved convincing; some have argued that the question as it stands, being too general, large, and vague, is unanswerable. However, this is not the place to review all these answers.[10]

Given its intractability, it is with trepidation that this author makes a stab to answer it. The brisk answer is this: China could never have developed what today we call "modern science," albeit European in origin but now globalized. To ask why China failed to develop modern science is to presuppose that it could, but for the intervention of certain factors that acted as impediments to such a progression. Examples of such impediments include the notion that China's cultural tradition looked down on merchants and entrepreneurs, while celebrating the literary scholarly class; its emphasis on bureaucratic administration at the expense of other intellectual pursuits; the nature of the Chinese language itself, and the lack in China of a monotheistic religion with a transcendent deity such as the Abrahamic one in Christianity.

The thesis proposed here to answer Needham's question suggests that the Chinese failure to develop modern science is not due *per se* to impediments such as the above, although some of the factors referred to could have played a role in discouraging scientific and technological pursuits in general.[11] More importantly the failure is primarily due to the different philosophical frameworks for science in China and Western Europe from the seventeenth century onward. Modern (Western) science with its crowning achievements in (Newtonian) physics and chemistry is a very specific distinctive kind of science with a distinctive methodology that cannot be understood independent of the metaphysics, the logic, as well as the causal paradigm it presupposes. By contrast, Chinese science in the form of CCM, adopted a metaphysics that entailed a methodology, logic, as well as concepts of cause, which are not compatible with the metaphysics and entailed methodology, logic, and concepts of cause in modern (Western) science. Hence, Chinese science often fails to be even recognized as "science" by people who have simply accepted that nothing counts as science unless it is an identikit instantiation of modern (Western) science.

CHINESE PHILOSOPHY AND CLASSICAL CHINESE MEDICINE

The differences between CCM and modern (Western) science illustrate why it is a grave mistake to divorce science from its philosophical underpinning, as if science is an activity that is free-floating without deep intellectual roots. Every intellectual endeavor (of which science is a preeminent example) is rooted in philosophy. Modern (Western) science is, therefore, rooted in modern (Western) philosophy.

However, if we envisage a different philosophy, we may also envisage a different science that such a philosophy could engender. In this spirit, let us first take a brief look at Chinese philosophy. To most people outside the Chinese tradition, the first thing that springs to mind when Chinese philosophy is mentioned would be Confucius and his school of thought that, of course, has profoundly shaped Chinese culture for more than two thousand years. As Confucius concentrated almost exclusively on social, moral, and political philosophy, one would be forgiven for thinking that Chinese philosophy consists of nothing but the value branch of philosophy.[12] This is not the place to discuss the diverse schools of ancient Chinese philosophy; however, since we are concerned about the relationship between Chinese philosophy and one possible instantiation of Chinese science in the form of CCM, it would make sense to focus on some philosophical aspects of two very different types of ancient Chinese texts—*Yi Jing or Zhouyi* commonly known in English as *I Ching*, and *Laozi*, also referred to as *Daode Jing*, which together with *Chuangzi* stand for Daoist philosophy.[13] The main canonical Daoist text considered here is *Daode Jing*, dated between 600 and 300 BCE, which is often attributed to Laozi himself.[14] Chinese scholarship considers this to have profoundly influence CCM.[15]

An older more difficult text, *Yi Jing*, contains a grouping of ideas that emerged as early as the ninth or eighth century BCE during the Western Zhou Dynasty.[16] Also pertinent is the fact that its main cosmological and metaphysical ideas were already extant by the time *Daode Jing* itself was written or compiled. In its long history *Yi Jing* came to be heavily used in connection with matters of divination, the very antithesis of scientific activities, although this is not to say that it began life as a book for divination. In antiquity it was used to predict the weather, but its use in meteorology should not be confused with plain divination. Indeed, *Yi Jing* in its basic earliest formulations set out certain cosmological principles at work that would give one an understanding of the natural processes at work in terms of the changes and movements of *qi*, which in turn would enable one to make predictions about the weather, as well as to understand the role played by such changes of *qi* in the lives of organisms, including humans. One can, therefore, readily distill from its cosmology and metaphysics certain key concepts that entered deeply into the formation of Chinese cultural and philosophical beliefs in general, and that which have held sway for several millennia, and shaped the formation of CCM.

Let us examine what are some of the main ideas common to these texts that provide the philosophical and cosmological foundation for CCM. These ideas are primarily metaphysical, including the concepts of *qi*, *wuxing*, and *yinyang* that are intimately intertwined. These are neither easy to translate nor put across in summary terms. However, *qi* has commonly been translated as "breath," "vital force," or "energy." *Qi* is cosmic in scope, permeating the universe (i.e., Heaven and Earth); it itself has no form though it manifests itself in all things, including therefore, humans as organisms and in all the activities they engage in as beings with self-consciousness and language, from the most sublime such as calligraphy or music, to the most mundane such as butchering. *Qi* originates ultimately from *Dao* (the Way).[17] This means that matter and energy are not two distinct substances—instead the one transforms itself into the other in a continuous ongoing process.

In the light of this understanding above, Western scholars have come to realize that it would be a mistake to translate *wuxing*—Metal, Wood, Water, Fire, and Earth—as five elements similar to the four elements of ancient Greek philosophy, namely, earth, water, air, and fire.[18] *Wuxing*, today, is translated as "five processes," or "five phases," emphasizing the dynamic character of this tradition of ancient Chinese philosophy. *Qi* may transform in any of the five phases of change and being.[19] Each process or phase can act upon another in two main cycles of balance, in reciprocal relationship to each other—the promotion cycle and the controlling cycle. Take the former cycle: spring qi initiates the process of rebirth in organic life and Wood is used to represent this phase when qi rises; Summer qi actively promotes the growth and flourishing of organic life and Fire is used to represent this phase when qi floats or rises to the surface; late or Indian summer qi refers to the stability reached by organic life and Earth is used to represent it; autumn qi is about the process of retraction as organic life begins to decline after reaching its peak of maturity and Metal is used to represent it when qi descends; winter qi refers to the process of retention, as organic life reaches the stage of rest or quietude and Water is used to represent it when qi sinks deep. The cycles proceed continuously, creating an endless flow between qi on the one hand and organic life on the other. Such a concept of change involves a non-Humean notion of reciprocal causality in which process A influences process B and process B in turn influences process A.

The notions of *yin and yang* also reveal the different structure of the logic of Chinese science. *Yang* refers to the space or time where or when the Sun shines, *yin* to the space or time in the absence of sunlight; more popularly, *yang* to the sunny side of the hill and *yin* to the shady side. *Yang* is said to be the active principle or male while *yin* represents the passive principle or female. However, *yinyang* should not be understood as similar to the dualism that modern Western philosophy, since Descartes, has celebrated. Dualism conceives of polarities in terms of a master and slave, or superior and inferior relationship, mind and body, human and nonhuman, culture and nature, male and female, light and dark, and so on. The first item in the pair is taken to be the master or superior class, the second to the slave or inferior class.[20] Hence, the first would be valued above the second. For example, it might be supposed that humans alone are superior and therefore have intrinsic value, while nonhuman others are inferior and, therefore, only have instrumental value for humans. But intrinsic to the concept of *yinyang* itself as a pair of polarities is that its framework is not dualist. *Yin* (passive) is not inferior to *yang* (active) since they are both essential to maintaining the dynamic equilibrium and harmony in the flow of qi in the cosmos and in the body. This kind of polarity is best understood in terms of dyadism[21] that recognizes a creative partnership and tension between them, without stigmatizing one as superior, and the other as inferior.

The metaphysics of the triad—*qi, wuxing,* and *yinyang*—in turn entails a logic that is not bivalent, but multivalent. Its values are not simply those of T (true) and F (false), or the logic gates of one and zero that are the surrogates of T and F of binary logic. In other words, it may be regarded as the forebear of what today is called "fuzzy" logic, a logic that has applications in diverse domains ranging from soft computing, control theory, as well as artificial intelligence.[22]

The famous *yinyang* logo shown in figure 11.1 displays vividly both its metaphysics and its logic.

Within *yang* may be found *yin*; within *yin* may be found *yang*; this means that they are not mutually exclusive and necessarily antagonistic. Even when *yang* is at its fullest, *yin* is there; even when *yin* is at its fullest *yang* is there. Furthermore, between *yin* and *yang* at their respective fullest, there are intermediary stages and processes where *yin* and *yang* may harmoniously be combined in certain proportions. Look at the eight trigrams in figure 11.2:

Figure 11.1   *Yin-yang* logo.

Figure 11.2   Eight trigrams.

The three unbroken lines show *yang* at its fullest, while the three broken lines show *yin* at its fullest. Summer is an instantiation of the former, yet even at the height of summer, as the *taiji* symbol of *yinyang* aptly shows, there is already the beginning of autumn. Furthermore, the trigram next to it clockwise has a broken bottom line, indicating that *yang* is on the decline and *yin* on the way to being ascendant, as summer turns into autumn. One could argue that the *yinyang*-cum-trigram combination above shows that *yinyang* logic displays the possibility of not merely two values, but six other values.[23]

In contrast to the *yinyang* logic of Chinese tradition, the logic of Western philosophy recognizes three core principles[24] first formulated systematically by Aristotle. They are the principle of identity (A is A), the principle of noncontradiction (A cannot be both B and not B), and the principle of the excluded middle (A is either B or not B).

Although *yinyang* logic appears not to challenge the principle of identity, it challenges the other two principles that are presupposed by binary logic. Where then do these principles stand with *yinyang* logic? To answer this question let us take the propositions:

1. "This glass is half full (p)"
2. "It is not the case that this glass is half full (–p)."

According to the principle of noncontradiction, p and –p cannot both be true at the same time. Yet according to *yinyang* logic, the glass may be said to be both half-full and not half-full at the same time. For instance, "the glass is half full" and "the glass is half empty" may both be true.[25] Indeed this move, which sidesteps the violation of the principle of noncontradiction, has been turned into a tautological truth, that is a definitional truth, by modern formal logicians.

Yinyang logic also appears to challenge the principle of the excluded middle. We have already seen that pure *yin* and pure *yang*—called the *kun gua* and the *qian gua* in the basic set of eight trigrams—constitute the polar extremes of the series of eight values. As binary logic permits only these two extreme values, it instantiates the principle of excluded middle—either pure *yin* is true or pure *yang* is true, but nothing in between. However, *yinyang* logic permits six other intermediate values. Using the example of the glass again: according to the principle of excluded middle, either "the glass is half empty" (p) is true or "it is not the case that the glass is half empty" (–p) is true. Hence it permits only p or –p to be true. However, according to *yinyang* logic, both p and –p may be true. Indeed it even permits that the glass could be less than half empty, more than half full, and so on.

Another way of putting the matter would be that the system of logic implicit in *yinyang* philosophy does not deny the principles of noncontradiction or of excluded middle *simpliciter*, but that it implies that these principles do not obtain in all contexts. They obtain only in certain limited contexts, namely at the polar extremes of the *yinyang* spectrum, but not in the other values indicated in the trigram. This would then be consistent with the claim that *yinyang* logic is a many-valued logic and that, therefore, it is a more comprehensive logic than binary logic that can then be seen as a subset within that wider many-value framework.

However it is important to emphasize that the plausibility of this interpretation of *yinyang* logic remains to be assessed. Also, it is important to note that the ancient Chinese philosophers themselves had not formalized a system of logic called *yinyang* logic, analogous to the system of binary logic established by Russell, Whitehead, and Frege. What the above argument suggests is that they implicitly used and invoked such a logic in their thinking and conceptualization of the world, and that today, should one wish to make it more explicit, then it could begin to look like a multivalent logic, of which "fuzzy" logic is a type.[26]

MODERN MEDICINE AND CLASSICAL CHINESE MEDICINE

Modern globalized medicine rests on modern (Western) science and its implied philosophy. As such, it takes on characteristics "dictated" by such a framework. Its notion of cause is Humean and linear,[27] and its logic is Aristotelian and bivalent. By contrast, CCM is ecosystemic in orientation with a reciprocal nonlinear notion of cause and its logic is multivalent. It assumes that the human being is an organism living within a larger environment that includes the physical and natural environment, the social and moral environment, as well as the wider cosmological environment of which the Sun and the Moon are part. Furthermore, the human organism must be understood in terms of its internal environments constituted by its various organ systems and how they reciprocally influence each other, as well as the reciprocal interactions between the human organism and the external environment.

We may elucidate the logic of CCM as follows: to be healthy, *yin* and *yang* in the human being must be in balance with each other. *Qi* would not flow without hindrance if there is excess or deficiency of *yin* or *yang*. For instance, too much *yang* especially at the wrong time of the year, such as in the winter when in terms of the natural environment *yang* is at its weakest, is a clear indication to the CCM diagnostician that the patient would be in a very bad way, if not dead, come the summer when external *yang* would exacerbate further his internal *yang*. *Wuxing* is incorporated as follows: for instance, liver is assigned to Wood, heart to Fire, kidney to Water.[28] Diagnosis and treatment must take into account the complex possible causal links between the five phases of change and being. Take the following example: when a patient presents himself to his doctor with insomnia, the doctor prescribes sleeping pills, which work upon the bio-chemistry of the brain sleep center, either by enhancing the sleep drive or by weakening the waking drive, targeting these with melatonin. CCM looks at the situation differently; there are different sources of insomnia, for which treatments involving different herbs are appropriate. One type is caused by a deficiency (*yin*) of kidney *qi*, such that the water of the kidney cannot rise to nourish the heart Fire, which then increases to disturb the spirit, thereby affecting sleep—in other words, the balance between kidney and heart is not in synchrony. CCM would recommend treating the kidney, which affects the heart that, in turn, enables the patient to overcome insomnia.

CONCLUSION

This chapter has argued that disagreement about substantial issues should not be settled through definitional maneuvers, and that which systems of thought and practice can legitimately be said to constitute a science should not depend on mere prior definitions of the term "science." It is a myth that any science is innocent and naïve of philosophy, in all its important dimensions of metaphysics, epistemology, methodology, logic, and cause. Modern science, resting on modern Western philosophy, has been very successful in understanding aspects of the physical world as well as manipulating and controlling that world for the past hundred and fifty years or more. However, in spite of its spectacular achievements, it has its weaknesses some of which may well stand in the way of our understanding the several crises facing our civilization in the twenty-first century. Therefore, it is time to turn our attention to an alternative paradigm of "post-postmodern" science that is in essence "ecosystemic" in orientation. This is the kind of perspective that we have argued under-pins CCM. Such a view also shows why The Needham Question appears to be misposed. The issue is not why the logic and causality of the ecosystems view came to be displaced by modern science, but why to resolve the contemporary environmental problems, we need to make modern science embrace the logic and causality of the ecosystems perspective.

END NOTES

1. See George Soros, "The Crisis and What to Do About It," *New York Review of Books*. December 4, 2008, 19; *The New Paradigm for Financial Markets: The Credit Crisis of 2008 and What It Means* (London: Public Affairs Ltd., 2008).

2. The term "modern" often refers to a philosophy engendering a new science initiated in Western Europe in the seventeenth century. By contrast, the term "postmodern" refers to a philosophy rejecting modernism, taking the

linguistic turnabout mid-twentieth century. Therefore, an approach that goes beyond modern philosophy and science, which has no sympathy with relativism, may be called "post-postmodern." However, as substantive issues do not depend on how one defines terms in language, those who do not like the term offered, are free to reject it.

3. We introduce the term "agent" here in a technical sense, namely, any entity or process that is capable of effecting change in any other entity or process.

4. These three may also be classified into two major groups: biotic agents (plant or animal) on the one hand, and abiotic (physical or chemical) agents on the other.

5. See Keekok Lee, *The Natural and the Artefactual: The Implications of Deep Science and Deep Technology for Environmental Philosophy* (Lanham, MD: Rowman & Littlefield, 1999).

6. The phenomenon is also called "global warming." These terms can be understood in one of two ways: that (1) Earth's temperature has risen over the last one hundred years or more; (2) this increase is anthropogenic. Global warming skeptics may be divided into groups—those who deny 2 but accept 1, and those who deny 1, thereby also denying 2 as an implication of 1.

7. Greenhouse gases have always existed in Earth's atmosphere, keeping it roughly at 33 degrees Celsius warmer than it would be without them. In their total absence, organisms would have not have evolved—organisms are carbon-based and would not normally thrive in the absence of warmth and light that come from the Sun. Greenhouse gases serve the useful function of retaining some of that radiation from the Sun while remitting the rest. It is their excess in the atmosphere that poses a threat. It is now acknowledged that global warming is alarming for the future of both humans and nonhumans.

8. Keekok Lee, *Social Philosophy and Ecological Scarcity* (London: Routledge, 1989, 70).

9. This case of nonlinear synergistic thinking is not meant to imply that the developments in mathematics and physics of late either cannot cope with nonlinear thinking or that they may not themselves be forms of nonlinear thinking. The point is simply to say that modern science since the seventeenth century is by and large Newtonian in conception, whose notion of cause is linear. That is why when quantum physics arrived on the scene, it raised and still raises numerous problems in any attempt, philosophically speaking, to reconcile these two very different paradigms of physics.

10. Joseph Needham, Kenneth Robinson, and Ray Huang, *Science and Technology 7*, part 2 (Cambridge: Cambridge University Press, 2004).

11. Confucian ideology, since its official adoption from the Han Dynasty, could have played an important role, as Confucian philosophy is primarily value-oriented with near exclusive preoccupation with the ethical, social, and political domain, shunning interest in the investigation of natural phenomena.

12. A recent major publication in English helps to reinforce this misleading view: Antonio S. Cua, ed. *Encyclopaedia of Chinese Philosophy* (London, Routledge, 2003).

13. This point, however, should not be interpreted to mean that no other types of philosophy, such as neo-Confucianism and Buddhism, had also influenced the development of Chinese medicine down the centuries.

14. This text is probably late fourth or third century BCE. Recent scholarship is based on two discoveries: two earlier silk scroll versions of the *Daode Jing* in the Mawangdui (Hunan province) excavation of a second BCE Han tomb in 1973; the *Guodian* bamboo slips (dated c. 300 BCE) found near the village Guodian in Hubei province in 1993.

15. One has to deal straightaway briefly with a potential criticism as some scholars, today, dispute whether one can so readily distinguish between Daoism as philosophy from Daoism as religion. However, religion, here, should not be understood in the Abrahamic sense. This matter is best left to the experts to debate. Suffice it to say that even though the two cannot be so readily distinguished in all contexts, it is still possible to distill from the Daoist lineage, concepts that are philosophical rather than religious in character. Chinese scholarship itself has always distinguished between Daoist philosophy (*Dao Jia*) from the Daoist religion (*Dao Jiao*).

16. See Richard J. Smith, *Fathoming the Cosmos and Ordering the World: The Yijing and Its Evolution in China* (Charlottesville: University of Virginia Press, 2008, 18–19). Recent scholarship of this text has also been greatly affected by the discovery in the Mawangdui excavation of a version earlier than any extant version. It throws doubt on the generally held belief, down the ages, that Confucius had a hand in editing Yi Jing.

17. A key passage of the *Daode Jing* reads: *The dao begot the one. One begot two. Two begot three. And three begot ten thousand things. The ten thousand things carry yin and embrace yang. They achieve harmony by balancing these forces.*

18. Sinologists in the past tended to be Classicists (whose original discipline was the study of Greek and Latin).

19. *Wuxing* is really short for *wu zhong liu xing zhi qi* that may be translated as "five kinds of *qi* dominating at different times."

20. See Val Plumwood, *Feminism and the Mastery of Nature* (London: Routledge, 1993).

21. See Lee, *The Natural and the Artefactual*, 1999.

22. The author would like to emphasize that what is said here and, indeed in the whole of this chapter should not be misunderstood to mean that the ancient Chinese argued explicitly for these ideas in the way set out here— suffice it to say that such drawing out on the part of the author is compatible and coherent with what is commonly attributed to the ancient Chinese texts cited here. On fuzzy logic, see: http://plato.stanford.edu/ entries/ logic-fuzzy/; Bart Kosko, *Fuzzy Thinking: The New Science of Fuzzy Logic* (London: Flamingo, 1994).

23. The schema can get more complicated. Later development of *Yi Jing* also puts two trigrams put together, creating first 16, then 32 but stopped at 64 hexagrams; these will be ignored for the purpose in hand, including a 12-value system called the *bie gua*, each value standing for the specific relationship between *yin* and *yang* for each month of the year.

24. Aristotle held them as regulative principles whereas classical logic holds them as tautological elements of a formal system.

25. One may try to argue that this is not a counter-example to the law of non-contradiction, as behind the two seemingly contradictory propositions stand the following true proposition: "this litre glass contains half a litre

of liquid." However, this simply shows that the reformulated version does not violate the law of identity, and not that as in the original version—"this glass is half empty" and "it is not the case that this glass is half empty"—are not both true at the same time. Another point that is very important to comment on briefly here is the usual charge that from contradictory premises any inference can be made. However, some logicians have begun to challenge this orthodoxy, through their work on paraconsistent logic. (I owe this point to Andrew Brennan.) See Graham Priest and Koji Tanaka. "Paraconsistent logic," http://plato.stanford.edu/entries/logic-paraconsistent/, 2009.

26. At the same time, it would be mistake to call it simply "informal," analogous to the informal logic advocated by "New Testament" Wittgenstein. Wittgenstein repudiated the main thesis of *Tractatus Philosophicus*, when he realized that ordinary language is not truth-functional, and that it does not correspond "one to one" with "world out there." However, instead of challenging binary logic, which underpinned the truth functional view of language, and developing an alternative that could do greater justice to ordinary language, he took the "linguistic turn" in *Philosophical Investigations*. By so doing, mainstream Western philosophy did not go down the route indicated by Max Black in his 1937 paper, "Vagueness: An Exercise in Logical Analysis," *Philosophy of Science* 4 (1937): 427–55.

27. There is an alternative model of cause in modern (Western) medicine, which this chapter calls "the ecosystemic paradigm," akin to that implied by CCM. It has been struggling to obtain *imprimatur* from the medical establishment (such as the Nobel Prize Committee for Medicine and Physiology) for the last one hundred years without success. It underpins epidemiological research; however, it is beyond the remit of this chapter to explore this matter further.

28. However, a word of caution is required here: An organ is not simply the organ that an anatomist can point to in the dead body lying on the marble slab in the pathology laboratory, but is to be understood as an organ system; neither can structure and function be distinctly separated out but must be understood in terms of the Janus-face aspects of an organ system.

12

DIVERSE CULTURAL CONTRIBUTIONS TO A "SCIENCE OF RELIGION": AN EMERGING ASIA-EUROPE DIALOGUE ON THE SCIENTIFIC STUDY OF RELIGION

Donald Wiebe

THE SCIENTIFIC STUDY OF RELIGION

The study of religion is, in one sense, as old as religion itself. In another sense, it is also considered to be a new enterprise—referred to variously as the science or the scientific study of religion, the history of religions, comparative religions, religiology, the academic study of religion, the nonconfessional study of religion, *Religionsforschung, Religionswissenschaft,* the rational exploration of religion, and, most commonly, religious studies—that only emerged in the West in the last half of the nineteenth century.

In its earliest forms the study of religion was clearly not a scientific undertaking. It was, rather, an activity concerned with the religious edification and spiritual growth of the individual and was, therefore, *primarily* a devotional enterprise. The objectives of such study of religion were, for the most part, training and formation in the practical and ethical requirements of the religious life. Insofar as being religious also concerned being able to provide a meaningful account of human existence, such study also involved considerable intellectual reflection in producing a coherent religious conception of the world. This kind of study, nevertheless, was essentially catechetical (confessional) in character rather than scholarly (nonconfessional). Most religious traditions, however, appear also to have inspired a more scholarly/systematic, and therefore in some sense scientific, investigation of "religious reality" and religious life that, even though still directed toward the edification of the believer, blended devotional and cognitive concerns in the quest for a more profound and shareable understanding of one's religious stance. In the Christian West, this kind of scholarship is referred to as *Glaubenswissenschaft* (a science of the Faith and/or a faith-imbued science); a form of scholarship that eventually produced an elaborate structure of theological disciplines that made possible the development of a more comprehensive, systematic, and therefore "scientific" (*wissenschaftliche*) understanding of the Christian faith. That is, it was scholarship from the point of view of the religious insider that produced a systematic body of "knowledge" about the Christian faith and worldview.

Such theological scholarship persists in many religious traditions today. However, there is no justification for considering it a form of the scientific study of religion. Even though much of the scholarly work in such disciplines as biblical studies and church history, for example, is indistinguishable from that produced by textual and historical scholars not constrained by a faith commitment and by religious belief, it is still an essentially religiously biased project given the broader university framework within which it is often incorporated. Being constrained by religious tradition and authority one might best, therefore, describe such a project simply as a religious study of religion.

The modern scientific study of religion, as it emerged in the West, however, is scientific (*wissenschaftlich*) in quite a different sense from that found in the notion of *Glaubenswissenschaft*, and is, therefore a new enterprise. This is not to say that religion is a *sui generis* reality and constitutes a theoretical object that justifies talk of a new discipline, but rather that such talk points to the development of an approach to the study of religion with an entirely different aim and a methodology wholly appropriate in the naturalistic framework of the modern research university in a way that *Glaubenswissenschaft* is not. The aim of this "new enterprise" is to obtain knowledge *about* religions and religious phenomena for the sake of that scientific knowledge alone, rather than, say, for the sake of consolation, for achieving spiritual illumination, or for laying down a foundation for a meaningful life. Its methodology, like that of the natural and social sciences in the context of the modern research university, functions on the assumption that the natural world is all there is and that it has mind-independent structures that can be discovered. In that sense, the scientific study of religion behaves, so to speak, as a naturalized citizen of the modern university and lives up to the same epistemic obligations as other scientific disciplines. Like them, it seeks "public knowledge" of "public facts"; it insists on the need for public accessibility to data, public intellectual operations that are subject to replication by others, and public testability of all hypotheses and theories so as to produce rationally warrantable claims (i.e., knowledge) about things, processes, and events in the world, including claims about human thought, behavior, and institutions. Scientific disciplines, moreover, have systematic and methodic procedures for reducing error in their "deliberations." In all these ways, the sciences exhibit a distinctive epistemic authority. They provide us with sets of mechanisms that actually explain the independent structures of the world assumed by naturalism and that justify Weber's claim that in science "no mysterious, incalculable forces come into play." It is true that science, and therefore the scientific study of religion, may be of little help in providing us with meaning or a sense of belonging in the world but that was not, nor is now, its primary objective; knowledge is. And in seeking knowledge *about* religion, religion *as* knowledge has been surpassed and the new study of religion (*Religionswissenschaft*) has found a legitimate place within the map of human studies.

THE SCIENCE OF RELIGION AS
A PRODUCT OF THE WEST

That the idea for a scientific study of religion of the kind described above emerged in the West is incontrovertible as Samuel Preus has shown in his book *Explaining Religion: Criticism and Theory from Bodin to Freud.* Drawing on Thomas Kuhn's notion of "paradigms," Preus identifies a series of minor revolutions in reflection on religion in the West, the cumulative effects of which produced a new intellectual ethos in which the study of religion was transformed from a religious to a scientific undertaking. That transformation, Preus maintains, began with the critique of religion by the sixteenth-century thinker Jean Bodin and found its completion with David Hume's attempt to account for religion nonreligiously. Bodin's analyses of Christianity, Judaism, and Islam, he argues, set off a progressively advancing series of criticism that, among intellectuals, resulted in an increasing detachment from religious presuppositions and commitments in attempting to understand religion despite religions' persistence in the population at large. David Hume, Preus argues, must be seen as the watershed between the traditional religious and theological understanding of religion and modern scientific accounts of it, for he "closes an era of criticism and opens the paths to future research" (100). "With him," writes Preus, "a line of criticism definitely ends and construction of alternative theory begins..." (84). Hume disposes of the last remaining religious explanations for, and legitimations of, religion and made possible other types of explanations that might be revealed in analysis of the connection between religion and the establishment and maintenance of social order, or the relationship between emotional and psychological needs of the individual.

Although there are earlier thinkers and scholars than Friedrich Max Müller who looked upon religion primarily as an object of study, it is

Müller who has received credit as a founding figure in the field because it is he who introduced the phrase "science of religion" into the discourse and insisted on the parallelism between the natural sciences and the study of religion. As Eric Sharp noted in his institutional history of the field, Müller's Hibbert Lectures, published in 1873 as *Introduction to the Science of Religion*, constitutes "the foundation document of comparative religion [the Science of Religion] in the English-speaking world" (1986, xi, 35). Although Müller neither abandoned his own religious concerns nor was negatively disposed toward them, he nevertheless was committed to a study of religion that rested wholly upon a scientific rather than a religio-theological motivation.

Though Müller did not provide a detailed program for the science of religion, he nevertheless reinforced the notion that being a scientific student of religion involves one in a set of activities that are clearly demarcated from the activities of the religious devotee. Although Müller held a belief in the ultimate convergence of religious and scientific thought and belief, he did not see this as implying that the scientific study of religion is, or ought to be, anything but scientific. In addition to the Hibbert Lectures, the "Essays on the Science of Religion" in the first volume of his *Chips from a German Workshop* and four volumes of Gifford Lectures make it clear that, for him, science is a search for knowledge about religion and that substantive religious truth claims, therefore, cannot constrain research programs in the study of religion. As he puts it in *Natural Religion*, the first volume of his Gifford Lectures, the same research program that has led to the success of the natural sciences "will be the most appropriate treatment of the Science of Religion..." and "we may readily accept the Science of Religion as one of the natural sciences in the most comprehensive meaning of that term" (1898, 13–14). It is within that general scientific framework that his theorizing about religion in relationship to the growth of language developed. In properly understanding the relationship between language and religion, he insisted, we will provide ourselves with "a truly historical basis for a scientific treatment of the principal religions of the world" (1898, 144).

Cornelis Petrus Tiele is another of the founding figures of the "Science of Religion." Although a professor of theology at Leiden University in the Netherlands, he worked assiduously to establish the study of religion as a scientific enterprise and, unlike Müller, was able to demarcate it from theology in a separate university department and provide for it the foundation of institutional structure.

Like Müller, Tiele seems to have held the belief that his Christian faith might find corroboration in the scientific study of religion, but he did not allow this hope to undermine his scientific intentions. Although it is clear that his religious convictions played some havoc with his work as a historian of religions, Tiele nevertheless exerted a positive influence on the development of a modern scientific approach to the study of religion within the university. In his two-volume Gifford Lectures entitled *Elements of a Science of Religion* he argued for an approach to understanding religion that can be ranked an independent science. And that required, he insisted, that the subject matter of this new science be available to all practitioners in the field. Thus, only "belief in the gods" calls for explanation, not the gods themselves. Consequently, Tiele defined religion not in terms of "the superhuman" but rather in terms of "the social institutions and customs" that produced *beliefs in the superhuman* that, unlike the gods, are available for examination, analysis, and explanation. As he put it: "The object of our science is not the superhuman itself, but religion based on belief in the superhuman; and the task of investigating religion as a historical, psychological, social, and wholly human phenomenon undoubtedly belongs to the domain of science" (1897, Vol.1, 5).

Although Müller was not successful in embedding the scientific study of religion in the curriculum of the university, as was Tiele, both were instrumental in creating another institutional structure for the support of this enterprise by participating in the first International Congress for the History of Religions. Although neither was present at that first meeting in Paris in 1900, due to illness, they consented to act as honorary presidents of the Congress. At this meeting it was agreed to establish an *ad hoc* committee to organize future congresses. This ultimately led to the formation of the International Association for the History of Religions (IAHR) in 1950. The thrust of this development is succinctly described by Eric Sharp in his history of the institutionalization of this enterprise in Europe, England, and North America: Since Paris, he writes:

> We have been able to see the field being gradually taken over by the advocates of comparative religion as a pure science. The irenic enthusiast was not welcomed, and soon came to realize that his interests would only be served by an entirely independent kind of gathering devoted to the goal of the final unity of all believers...[B]y the 1920s the two paths [of the study of religion as a pure science and as an applied science] had become almost entirely separated. (1986, 252)

THE SCIENCE OF RELIGION AND LOCAL KNOWLEDGES

My account here of the nature and history of the development of the scientific study of religion shows that it is a natural outcome of the European Enlightenment and the rise of modern science. Whether the emergence of such an approach to understanding religious phenomena is a good thing, and whether those developments are necessary preconditions for

226 DONALD WIEBE

a "Science of Religion," are matters not settled by my account. Although, I think it cannot be denied that the study of religion as a scientific enterprise does now exist (though often blended with traditional religious motivation and methodologies) and that it was first legitimated as such by its incorporation into the modern European university, it is the case nevertheless that there are scholars who maintain that it is a cliché to talk about a scientific study of religion as the product of a Western approach to knowledge and a child of the European Enlightenment. They believe, rather, that the cliché has spawned an illegitimate and harmful new "epistemic regime" in the study of religion that continues to operate from within a colonialist framework. I have in mind here David Chidester's *Savage Systems: Colonialism and Comparative Religions in Southern Africa* (1996), Richard King's *Orientalism and Religion: Postcolonial Theory, India, and the "Mystic East"* (1999), and Gregory Alles's "Cargo Cult Science and the Study of Religions: Genealogy in an Age of Globalization" (2008a).

"Colonialism and Comparative Religions in Southern Africa"

According to Chidester all disciplinary histories are invented (xii) and the Eurocentric history of the study of religions simply does not tell "the real story" behind the rise of religious studies (that is, the academically legitimated scientific study of religions). He attempts to argue that it is not the desire to obtain knowledge but rather the lust for power that underlies the academic study of religion that we see in the modern university. According to him, religious studies in South Africa was a "frontier enterprise" carried on by travelers, traders, colonial agents, other government officials, and indigenous peoples, and that the product of their combined labor constituted "a human science with its techniques of observation, conversation, and description, and its procedures of analogical reasoning and theoretical generalization" (1). In the process, he maintains, these European observers generated "knowledge" about religion that would aid them in controlling the indigenous population in various regions of southern Africa. Chidester also maintains that this "knowledge" played an important role in the development of the academic study of religion in Europe; that European scholars appropriated the "data" of these "comparativisms" in the building of their "science of religion," thus making that science an imperialist enterprise embroiled in the production of "global knowledge" that "reinforced power relations over colonialized peripheries" (28).

I find Chidester's argument unpersuasive given (1) the internal inconsistencies in his alternative history of the field, (2) his assumption that all those ready to speak about religion in South Africa must be recognized as comparative religionists, and (3) his assumption that reliance by European scholars on the data from informants in the field is always uncritical and always constitutes an espousal of the ideological frameworks within which these so-called alternative comparativists worked. Chidester seems unable to recognize that the scientific study of religion emerging in the universities of Europe was self-consciously structured to be critically alert to the possible contamination of data by political bias and sensitive to the possibility of ideological distortion of "the facts." Quite simply, Chidester has not shown a clear developmental connection between what he calls the history of comparison within the colonial conditions of southern Africa and the emergence and development of the scientific study of religion in the academic centers of Europe and North America.

"Postcolonial Theory, India, and the 'Mystic East'"

Richard King's critique of the study of religion conceived scientifically, like that of Chidester, draws heavily on the ideas of Foucault and Said. Although he acknowledges that the European Enlightenment made possible a study of religion as a new, nonconfessional, secular discipline, and that as such, it "has clear advantages in terms of the politics of the institution and prevailing attitudes towards the academic work of the department," (42) he, paradoxically, also believes the description of the study of religion as scientific to be a misrepresentation of what it really is. He insists that the so-called scientific study of religion cannot lay claim to being objective and without bias because it is itself an aggressive, antireligious worldview—an "irreligious dogmatism of secular reductionism"—given its espousal of naturalism. The modern scientific study of religion that I have described above, therefore, is for King a "regime of epistemic violence" in that it erases the religious dimension of religious phenomena and is seriously entangled in the imperialism of the West.

I do not find King's argument cogent. He admits the emergence of a new, scientific way of approaching the study of religion, but at the same time denies that it can actually provide us with any knowledge about religion because it is nothing more than an irreligious rejection of religion, and students of the enterprise are "little more than a priesthood for militant secularism" (48). But all of this is little more than gratuitous namecalling. Furthermore, King proposes an alternative approach to the study of religion that he names "cultural studies," an account of which seems to do little more than recapitulate the empirical and descriptive aspect of the scientific study of religion described above. His "cultural studies" approach involves nothing more than sensitivity to the complex environment in which religious phenomena are found and an awareness of the possibility of bias playing into its interpretation. His proposal here, however, shows a failure to understand what the scientific study of religion is all about, which is, ultimately, to seek explanations and theories to account for religious phenomena only after it has gained what King calls "a rounded account" and an "insightful understanding" of the phenomenon being examined (59, 80). Finally, King's claim that because some European academics have been seriously entangled in imperialist projects of the West, the Western study of religion is therefore a "regime of epistemic violence" is a *non-sequitur*; nor does he provide any justification of or any evidence for the implied claim that the scientific study of religion is necessarily imperialist in intention or action.

"Cargo Cult Science and the Study of Religions"

In his essay on cargo cults and the study of religion, Gregory Alles argues against the claim that the scientific study of religion is a natural product of the European Enlightenment; a claim he considers an empty cliché. A proper understanding of the field of religious studies, he claims, requires a global rather than just a European history. He argues that there are forms of the study of religion that are indigenous counterparts to the Western, scientific study of religion that should allow scholars in every culture to identify scholarly ancestors among their fellows (21). According to Alles, for example, we do cargo cultists a disservice if we see them only as religious leaders and not also as scholars of religion (25, 29). Speaking of a particular cargo cultist named Yali he writes: "Virtually every intellectual move that I have attributed to Yali...has its counterpart in the study of religions by Europeans and North Americans" (28). This seems to me a strange claim, but Alles sees it as justifiable if one views science not in terms of institutionalized structures but rather in terms of investigative procedures or, as he puts it, as an enterprise "consisting of critically scrutinized observations about the world, combined with attempts both to postulate mechanisms that account for what critical observation reveals and to refine those mechanisms by trying to determine whether and where they fail" (29). What Alles does not consider, however, is other "moves" that cargo cultists make and modern Western scientists do not. He admits, for example, that the cultists often affirm bizarre statements, but brushes this off with the comment that some of the statements Europeans and North Americans make are also bizarre. The issue, however, is whether such statements are as pervasive in scientific accounts of events as they are in cultist accounts. Alles, unfortunately, does not ask that question. He simply claims that cultists work within certain cognitive constraints but fails to show that they are anywhere near the same constraints that are imposed by the naturalist framework within which the Western scientific student of religion operates and this, I think, undermines the basic thrust of his argument.

In summary, then, I find Alles's willingness to consider that "cargo cults [are] a kind of indigenous counterpart to anthropology," and his suggestion "that cargo cults are forms of knowledge of religion," that justifies taking cargo cultists as "disciplinary ancestors" (22) unacceptable. There may be some justification for the claim if it is restricted to "ancestors" in the sense of critics of the cult, or the work of "cultist intellectuals" who produce a kind of *Glaubenswissenschaft* that makes use of critical intellectual tools to better understand the religious world view in question. However, unless a wholly alternative framework (free from the religious concerns for meaning, value, and truth) for the understanding and explanation of religion is created, it is difficult to see any ground for the claims Alles makes. Alles is wrong in my judgment, therefore, to claim that the agenda of Yali and other cargo cultists makes them "not theologians but in some sense scholars of religion" (29).

Parallel Developments in Other Cultural Contexts

I agree with Alles that "the ambition to formulate a definitive genealogy for the study of religion, if anyone has such an ambition, is a misplaced dream" (20). Although it does seem that the prevalence of the new non-confessional approach to the study of religion in the West has much to do with the scientific revolution and the *intellectual* and political ethos in modern Europe, I do not mean to suggest that the account that I have provided of its emergence and development constitutes a definitive genealogy of the "science of religion." Nor should my rejection of the critiques of the notion of the scientific study of religions be read as support for such a genealogy. Indeed, in this section of the chapter, I will attempt to show the credibility of that so-called Western notion by providing evidence of at least the possibility for the emergence of such a nonconfessional and scientific study of religion in non-Western and non-Christian cultural contexts.

Islamic Precursors to a Science of Religion

In summarizing a research project presented to a conference on the topic of "Institutions and Strategies in the Study of Religion" two decades ago, Peter Antes claimed "that Islam has had a tradition of religious studies long before those studies started in Europe" (1989, 148). "Famous Muslim scholars in the past," he writes, "have already provided an example of religious studies as understood today in the International Association for the History of Religions" (143). Abrahim H. Khan in his essay on "The Academic Study of Religion With Reference to Islam" (1990) also argues that in tenth-century Baghdad there existed a form of naturalism in the work of thinkers such as Ibn ar-Rāwandī (d. ca. 910) and ar-Rāzī (d. ca. 932) that opened up religious faith to examination. The work of al-Bērūnī (d. 1048) a century later, he points out, was intended to provide an objective account of the religious beliefs of India and that his work should be considered to be the first work in comparative religion (1990, 39). Khan refers to these developments as "fleeting moments of intellectual boldness" that indicate the evidence of "a spirit of free enquiry" (39) but claims that they never really "developed into the academic study of religion" (40). There were other such experiments in subsequent centuries that also showed promise for the creation of a scientific study of religion. Ibn Khaldūn (1332–1406), he suggests, advanced "the scientific study of culture along positivist lines...[providing] explanations of historical and cultural processes in terms of causal connections, secular principles, and general laws rather than with reference to belief in divine providence or moral retribution" (40). Khan sees the work of Sayyd Ahmed Khān (1817–1898) as yet another opportunity for the emergence of a science of religion. About Khān he writes: "He made natural reason the norm for determining what will count in his reorganizing of Islamic beliefs to overcome retrogressive attitudes or impediments to cultural progress" (40).

In a recent treatment of the study of religion in North Africa and West Asia in a volume on *Religious Studies: A Global View*, Patrice Brodeur also speaks of a period of a protoscientific study of religion in a Muslim context. According to Brodeur, this occurred in the golden age of Arabic literature that permitted a certain "methodological self-reflexivity" (82) that he believes can be spoken of as the origins of a scientific study of religion (87). "The fifth Islamic century," he writes, "came to a close with the production of the first work on religious others composed in the Persian language...written in 1092 C.E. by Abū al-Maʿālī al Alavī" (84). He also points out that al-Sharastānī's (1076–1153) *Book of Religious Communities and Systems of Thought* is often seen by scholars as the first genuine history of religions (85). Brodeur points out, however, that these protoscientific works served not only the intellectual community but also political elites and this has an important effect on the scholarship involved because it was constrained to some extent by the need to provide a collective identity for the group by "solidfying the boundaries of Muslim identity over and against a set of distinguishable others that had clearly been subjected by them to political domination" (87). Finally, by the seventeenth century there were no further considerations of religious others.

The various protoscientific studies of religion to be found in Islamic contexts, then, clearly indicate the potential for a science of religion as described in the first section of this essay, even though such a science did not fully develop in Muslim societies. Antes maintains simply that "these studies have not been renewed in more recent times" because of "political circumstances" (156). "Fears of seeing Islam attacked," he continues, have "hindered many scholars in the Muslim world from participating in the debates as equal partners with comparable works of their own" (156). A further contributing factor to the decline in such studies, says Antes, is institutional, in that the study of religion in the Muslim world was carried on in madrassas that concerned themselves with "a totalized vision of life and learning" rather than an institution comparable to the modern Western university. There is a study of religion today in Islamic countries within a framework of interreligious dialogue which he claims, however, is not that of the IAHR. Abrahim Khan also maintains that the moments of intellectual boldness in early writings about religion in Islam never developed into the academic study of religion because of fears that this might amount to an attack on Islam; that it would "suffocate Islam" (40). Khan also claims that there was simply no motivation to study other religions. On the one hand, priority was given "to Muslim religious sciences" rather than to the intellectual or "foreign ones," and, on the other, "the theological underpinnings of [Islamic] culture were such that the study of other cultures was irrelevant"; they were simply not considered worthy of study (43).

Although Brodeur questions the assumptions underlying what he calls Western religious studies (refusing to identify the scientific study of religion negatively as nonconfessional or nontheological [87, 90]) he does admit that the kind of science of religion that I have described above emerged "with the development of various initially European modernities" (88), and was later exported to other regions in the world. "At the heart of these modernities," he writes, "lay the powerful positivistic scientific discourses out of which emerged a 'science of religion' (*Religionswissenschaft*) or 'sciences of religion,' in opposition to the pre-modern medieval theological discourses, most in their European Catholic and Protestant Christian expressions" (88). But the institutional structures that can support such an enterprise, he goes on to argue, depend upon certain economic underpinnings and political resources such as a democratic nation-state or near equivalent. Where a science of religion has flourished in modern Muslim societies, therefore, it has done so in premodern educational institutions that have undergone considerable transformation, or in universities with roots in missionary institutions, or in national universities in newly independent nation-states, or, finally, in private institutions of higher learning (90–98). However, as Jacques Waardenburg points out in "Observations on the Scholarly Study of Religions As Pursued in Muslim Countries" (2000), "the field of Religious Studies or History of Religions has not yet been institutionalized" (98) and he advises those who wish to study religions and cultures scientifically to enroll in Western universities (102).

Proto-Religionswissenschaft in
East Asian Cultures

In a near score of articles over a period of three decades Michael Pye has forcefully argued that "the emergence of a science of religion is not exclusively linked to a set of postulates available only in western culture, as is often supposed" (1992, 108). In an article on the understanding of the notion of religion in East Asia he writes: "There is no need to espouse a culturally isolationist view of the history of the development of reflection about religion [because the] reality is that abstract, critical reflection about that which in western languages we call 'religion' emerged both in Europe and in East Asia even though the terminology has, of course, varied" (1994, 115). The term "religion," he insists therefore, cannot simply be written off as a misleading Western import into East Asian thought. In arguing his case, Pye points out, for example, that as early as the fourteenth century, long before any influence of the West on China, the traditions of Confucianism, Buddhism, and Taoism had been "the object[s] of sustained historical and systematic reflection" in Ming Tai Tsu's effort to bring about religious stability in society. In an essay on "Three Teachings (*sanjiao*) Theory and Modern Reflection on Religion," Pye argues that this process of critical reflection about the plurality of religions was undertaken in a dispassionate way as social systems that needed evaluation. Moreover such analysis, Pye maintains, made room for critical reflection on religions free from religious control or influence and in that process identified religion as a phenomenon that required explanation. Consequently, according to Pye, reflection on the plurality of religions in China was as much a starting point for a scientific study of religion in the East as it was in the West. As Pye puts it, "recognition of religions in their plurality is a fundamental frame of reference in terms of which the study of religion can be carried out" (1995, 116). Indeed, in a later essay on "East Asian Rationality and the Exploration of Religion" (1997a), he insists that when people are forced to consider the issue of the plurality of religions the conditions for the emergence of a proto-*Religionswissenschaft*, if not *Religionswissenschaft* proper, exist. And according to Pye, one can find rational reflection on religion not only in China but also in India, Korea, Vietnam, and Thailand, indicating that what he calls "the rational exploration of religion" must be seen as "a shared intellectual enterprise" with far more starting points than is generally recognized by the West. Indeed, for Pye, knowledge of these developments calls for rewriting the story of the emergence and development of the rational or scientific study of religion. "In the future," he writes, alluding to Eric Sharp's Western-oriented history of the field of religious studies, "a history of the history of religions will not be restricted to Europe and North America, but will have to take several, diverse developments into account" (1997a, 76).

Pye first recognized this need for a drastic rejection of the widespread assumption that the emergence of a scientific approach to the study of religion is a wholly European phenomenon, in his discovery of the work of Nakamoto Tominaga. In "Aufklärung and Religion in Europe," he compares Tominaga's work with that of Gotthold Ephraim Lessing and claims that we see in such a comparison "that western thought about religion is not unique" (1973, 203, n3), for it shows that Tominaga was not inspired by religious apologetics or reformism but rather that he tried to understand religion in terms of its chronological and cultural context. In a later essay, Pye points out that Tominaga moved from a critique of religion to a search for general laws of religious development. As he puts it, there was "a sustained effort of reasoned enquiry in the Tokugawa period which included a very modern-looking, historical critical theory of religious tradition" (1982, 16) within which Tominaga worked. Pye also claims in a paper delivered to the fourteenth International Congress of the IAHR that Tominaga's work *parallels* what was done in the West (1983, 566) and insists, therefore, that it cannot be argued that European thought is alien to the Asian intellectual tradition. "The plain fact is," he writes, "that in many important respects Japanese thinkers had thought in modern modes long before they were called upon to modernize" (1983, 576). Thus, he writes in the preface to his translation of Tominaga's *Emerging from Meditation*, they were "able to assimilate and respond with tremendous speed and skill" to Western influence (1990, 1). And he goes on to say there that Tominaga's work is sufficient evidence for positing "the universal appropriateness of historical and theoretical inquiry into religion without fearing any accusation of cultural imperialism" (1990, 14).

In light of Tominaga's work in Japan, and similar developments in other East Asian contexts, there can be no excuse for the ignorance in European and North American circles for the intercultural location of the religious studies enterprise. This is particularly so regarding Tominaga's work, Pye insists, because it "has been followed through, after some fits and starts and with significant western stimulus in the nineteenth century, to the point where now a major non-western discipline of religious studies exists in Japan" (1983, 565). Indeed, the Japanese Society for the Study of Religion, established in the 1930s, is the largest national or regional member organization in the IAHR. And Pye notes that the "conversational partnerships between the Japanese and western based study of religion" offers a real chance for the subject not to be determined by a single set of cultural perspectives, but offers instead to develop for it a more widely valid intellectual profile.

Although Pye maintains that the work of Tominaga, and other East Asian scholars more generally, provided what he calls "a transcultural base for developing the historical and comparative study of religion" (1983, 576), he is quite aware that the rational study of religion is not carried out in the East in the same way that it is carried out in Europe and North America. Religious studies in East Asian countries do not in his judgment constitute a fully scientific study of religion as he notes in his essay on the 'three teachings theory' referred to above (1995, 116). Elsewhere he also claims that the scientific approach to understanding religion in East Asian countries appeared only "fitfully" (1997b, 9), or merely in "incipient form" (2003, 112). In his essay on "An Asian Starting Point for the Study of Religion," he expresses these concerns in greater detail. He writes:

> Admittedly the detached, analytical study of religion is of marginal signifi-
> cance in most cultures, even if it appears at all, and it is particularly frag-
> ile where religion itself is a dominant force. As a result, Islamic Studies,
> Buddhist Studies, and the like, tend to become established rather than
> an autonomous, historical, comparative, or phenomenological discipline.
> Such a scientific study of religion may be regarded as potentially mislead-
> ing or even hostile to religion. (1992, 102)

CONCLUSION: AN EMERGING
ASIA-EUROPE DIALOGUE

In his book on *The Dialogue of Civilizations in the Birth of Modern Science,* Arun Bala argues persuasively that it was Europe's receptivity to ideas from beyond its geopolitical boundaries, and its ability "to trans-form and assimilate them" (25), that made possible the emergence of modern science. This "dialogical contribution," as he calls it—of which most moderns, he points out, have little knowledge—does not under-mine the claim that the emergence of modern science involved a revo-lutionary break. Bala writes: "It is hardly ethnocentric to claim that a scientific revolution occurred in Europe—unless one denies the plausible claim that modern science first emerged in Europe" (38).

I think it just as plausible to claim that the modern scientific study of religion first emerged in Europe and that it did so largely because of the work of a few "religious scholars" who were, so to speak, in the thrall of modern science and believed that it is possible to harmonize what they thought of as religious and scientific "truth." Although there were philo-sophical schools of thought in non-Western cultural contexts that made possible the *critical assessment* of religious movements and traditions, there is no evidence that they supported the emergence of a fully blown science of religion or played a significant role in the emergence of the science of religion in the West. That is, we have no indication of the influence of "alternative cognitive traditions" and "global traditions of science," on the emergence of the science of religion in the West of the sort that Bala sees as having influenced the emergence of the modern natural sciences in Europe (4, 176). There is a sense, therefore, in saying that the emergence of the modern science of religion is a product of European thought and that it emerged independently of any direct "dialogical contribution," although it subsequently became aware of the possibilities for such a science in other cultural contexts. That is, the critical mindset that emerged in Europe and that grounded a scientific approach to understanding religion also existed in many non-Western cultures. As Pye points out, "the European Enlightenment is simply not alien...to the Asian intellectual tradition" (1983, 565).

Given the existence of that similarity in Eastern and Western intellectual traditions, and the incipient formation of nonconfessional and nonreligious approaches to accounting for religious plurality in non-Western cultures, it is obvious that critics such as Richard King are unjustified in claiming that "religious studies" constitutes a "regime of epistemic violence" (1999, 199) that amounts to little more than a "systematic and violent suppression of non-Western ways of life, forms of knowledge and constructions of reality" (187). The evidence provided indicates that the term "religion" can't be written off as a misleading Western import and that the "thrust towards the rational exploration of religion," as Pye puts it (1997a, 67) is not a European construct that is imposed upon other civilizations. There is, therefore, no basis for an accusation of cultural imperialism against the science of religion (1989, 14).

The evidence clearly indicates that the modern science of religion is much more solidly entrenched in Europe and North America than it is in the East, even though Pye is entirely on the mark in his comment that the general perception in the East of such a study of religion being misleading or even hostile to religion is a reaction that "is not exactly unknown in the western world..." (1992, 102). Nevertheless, the evidence also suggests that there is a solid transcultural base upon which a global historical, comparative, and theoretical study of religion can be established, as does the preliminary "mapping" of work in religious studies being done around the world in the volume *Religious Studies: A Global View* (2008b) edited by Gregory Alles. Given the importance of the institutionalization of such intellectual and academic enterprises to their long-term success, it seems to me that the intercultural conversation that needs to take place in this regard might best be facilitated by organizations that distinguish the objectives of the study of religion from religious and other more-than-epistemic goals. This is not to say that the results of such study are necessarily irrelevant to nonepistemic goals such as theological reflection or interreligious dialogue. But these are not matters under consideration in this context.

BIBLIOGRAPHY

Alles, Gregory. 2008a. "Cargo Cult Science and the Study of Religions: Genealogy in an Age of Globalization." In *Introducing Religion: Essays in Honour of Jonathan Z. Smith*, edited by Russell McCutcheon and Willi Braun pp. 18–39. London: Equinox Press.
———. 2008b. *Religious Studies: A Global View*. London: Routledge.
Antes, Peter. 1989. "Religious Studies in the Context of Islamic Culture" In *Marburg Revisited*, edited by Michael Pye, pp. 143–156. Marburg: diagonal-Verlag.
Bala, Arun. 2006. *The Dialogue of Civilizations in the Birth of Modern Science*. New York: Palgrave Press.
Brodeur, Patrice. 2008. "North Africa and West Asia." In *Religious Studies: A Global View*, edited by Gregory Alles, pp. 75–101. London: Routledge.
Chidester, David. 1996. *Savage Systems: Colonialism and Comparative Religions in Southern Africa*. Charlottsville: University of Virginia Press.
Khan, Abraham H. 1990."The Academic Study of Religion with Reference to Islam." *Scottish Journal of Religious Studies* 11, no.1: 37–46.
King, Richard. 1999. *Orientalism and Religion: Postcolonial Theory, India, and the "Mythic East."* London: Routledge.
Müller, Friedrich Max. 1898 [1889]. *Natural Religion*. London: Longmans, Green, and Co.
———. 1881 [1856]. *Essays on the Science of Religion, Vol 1, Chips from a German Workshop*. New York: Charles Scribner's Sons.
———. 1873. *Introduction to the Science of Religion*. London: Longmans, Green, and Co.
Preus, J. Samuel. 1987. *Explaining Religion: Criticism and Theory from Bodin to Freud*. New Haven, CT: Yale University Press.
Pye, Michael. 2003. "Overcoming Westernism: The End of Orientalism and Occidentalism." In *Religion im Spiegelkabinett: Asiatische Religionsgeschichte im Spannunbsveld zwieschen Orientalismus und Okzidentalismus*, edited by Peter Shalk, pp. 91–114. Stockholm: Elanders Gotab.
———. 1997a. "East Asian Rationality in the Exploration of Religion." In *Rationality and the Study of Religion*, edited by J. S. Jenson and L. H. Martin, pp. 65–77. Aarhus: Aarhus University Press.
———. 1997b. "Reflecting on the Plurality of Religions." *Marburg Journal of Religion* 2, no. 1: 1–11.
———. 1995. "Three Teachings (*sanjiao*) Theory and Modern Reflection on Religion." In *Religion and Modernization in China*, edited by Dai Kargsheng, Shang Xinying, and Michael Pye, pp. 111–116. Cambridge: Roots and Branches Press.
———. 1994. "What is 'Religion' in East Asia?" In *The Notion of "Religion" in Comparative Research (Selected Proceedings of the XVIth International*

Association for the History of Religions Congress), edited by Ugo Bianchi, pp. 115–22. Rome: L'Erma Di Bretschneider.

———. 1992. "An Asian Starting Point for the Study of Religion." In *Religion in History: The Word, the Idea, the Reality*, pp. 101–109. Waterloo, ON: Wilfred Laurier University Press.

———. 1990. "Introduction." In Tominaga Nakamoto, *Emerging from Meditation*, edited and translated by Michael Pye, pp. 1–47. Honolulu: University of Hawaii Press.

———. 1989. "Cultural and Organizational Perspectives in the Study of Religion." In *Marburg Revisited: Institutions and Strategies in the Study of Religion*, edited by Michael Pye, pp. 11–17. Marburg: Diagonal-Verlag.

———. 1983. "The Significance of the Japanese Intellectual Tradition for the History of Religions." In *Traditions in Contact and Change: Selected Proceedings of the XIVth Congress of the International Association for the History of Religions*, edited by Peter Slater and Donald Wiebe, pp. 565–577. Waterloo, ON: Wilfred Laurier University Press.

———. 1982. "Religion and Reason in the Japanese Experience." *King's Theological Review* 15: 14–17.

———. 1973. "Aufklärung and Religion in Europe and Japan." *Religious Studies* 9, no. 2: 201–217.

Sharp, Eric. 1986 [1975]. *Comparative Religion: A History*. London: Duckworth.

Tiele, Cornelis Petrus. 1897. *Elements of a Science of Religion*, 2 vols. London: William Blackwood and Sons.

Waardenburg, Jacques. 2000. "Observations on the Scholarly Study of Religions As Pursued in Muslim Countries." In *Perspectives on Method and Theory in the Study of Religion* edited by Armin Geertz and Russell McCutcheon, pp. 99–109. Leiden: E. J. Brill.

Weber, Max. 1946 [1919]. "Science as a Vocation." In *From Max Weber: Essays in Sociology*, edited and translated by H. H. Gerth and G. Wright Mills, pp. 129–156. Oxford: Oxford University Press.

13

RECLAIMING TRADITION: IMPLICATIONS OF A KNOWLEDGE INDIGENIZATION PERSPECTIVE FOR ASIAN EDUCATION

S. Gopinathan

While there is a considerable body of research and disputation in the history of science, and the unresolved questions therein, there is little of a similar nature in the education sciences. In some ways, the state of play in the education sciences perhaps represents that in the history of science several decades ago. Just as some scholars are convinced that the rise of science in the West owes little to Eastern science, there is, for instance, among educationists of a broad stripe and in institutions such as the World Bank, a belief—even conviction—that Western models of education, more specifically schooling, are exactly what is needed for post-colonial societies wanting to modernize. Education libraries are full of books on the processes, history, and philosophy of education that make few, if any, references to non-Western traditions. Other examples are the privileging of the use of metropolitan and colonial languages in education; the strong desire to study abroad and earn recognition by acquiring foreign qualifications; and the often uncritical importation of Western, largely US education, trends and initiatives in education policy and practice. It is not difficult, for instance, to find enthusiasm for such schemes as vouchers and public-private partnerships in school management and governance in the Gulf states in the Middle East. More recently, there has been much excitement among policy elites in developing countries about evidence-based policy and pedagogic practices, unaccompanied by any understanding of the social and political contexts in the United States, from which these notions largely originated.

Much of what characterizes modern-day education in Asian societies is almost a mirror image of those in the West. Many aspects, such as the philosophy of early childhood education, are drawn from the insights of Rousseau, Montessori, Piaget, among others. While many systems now use indigenous languages as media of instruction and all attempt to teach English as a second or foreign language, there continue to be ongoing disputes, as in Malaysia, over language-of-instruction and assessment issues a half-century after independence. Elites, in almost all parts of the developing world, send their children to private schools in which English is widely used. These schools use examinations conducted by boards located in the United Kingdom and Europe. While there has been some indigenization of the curriculum, far more attention continues to be given to events and aspects in say, European history, than in Asian history. Primary and secondary divisions, the influence of the Cambridge Examinations Syndicate, the Scholastic Aptitude Test, and now the International Baccalaureate, all bear testimony to the dominance of the West. A Western model of schooling and postsecondary education, it can be argued, reigns unchallenged in the East.

While much of this is due to colonialism, even nations that were not colonized, such as Thailand, have adopted this model. Japan learned from Western best practices in its historic attempt to modernize. Western firepower, the attractions of industrialization, primarily its capacity to create wealth, and soft power like the Christian faith all played a part. In Singapore, for instance, the Catholic, Methodist, and Anglican churches, all founded very successful English-medium schools with the encouragement and support of the colonial government. Though Singapore has made an attempt to retain some traditional features of Chinese clan-founded schools, little that is traditional remains. Even the Malay madrassas are being asked to modernize and adopt state-sponsored curricula and examinations.

The Needham question for the history of science can be rephrased for education in the following terms:

1. What are the achievements and traditions in education in Eastern societies?
2. What contributions did these traditions make to the cultural and scientific achievements of Eastern societies?
3. When and how did they fall into decline and what was the role of Western colonial activity in this decline?
4. What are the critical factors/processes that led to the imposition/adoption in the colonial and postcolonial eras of mass modern state-sponsored schooling?
5. What were the ways in which education ideas and structures crossed cultural boundaries and what impact did they have?
6. What can we learn from educational cross-cultural dialogues that can help us better understand how we might strengthen education across cultures?

An argument can be made that more attention should be paid, especially at this time, to educational processes, especially in the context of efforts to promote intercultural dialogue. Schools are after all primary sites for socialization and in many multicultural societies are vital to the processes of knowledge- and disposition-building that is essential for social cohesion. Yet, even in many multiethnic and cultural societies, now open to the globalization process, the efforts are nation-centric. In more settled nations with a stronger cultural identity such as Indonesia or Thailand there is a stronger national presence in textbooks; in others there is still too little attention to the contributions of Asian societies, thinkers, and innovators.

As noted earlier, the questions to ask are: (1) why has this happened? (2) What consequences flow from it? and (3) What can be done about it? Trade and cultural contacts between Eastern and Western societies is a long process, going back in Peninsular Malaya to the sixteenth century. While it is acknowledged that such early contact between cultures and civilizations must have led to exchanges and transfers of knowledge, the focus here is in the more intense and transformational imperialism process from the mid-eighteenth century. It was during this period that the political, economic, and sociocultural domains in Asia were most impacted upon.

It has to be noted at the outset that there is certainly evidence of an increasing disquiet over this state of affairs in the education sciences. This is due in part to the "rediscovery" of the complexity and richness of indigenous cultures, ironically, during the colonial period by officials of say, the East India Company and the colonial administrative services—men such as Raffles and Swettenham in Malaya and in India, scholar-administrators such as William Jones who translated Kalidasa, James Prinsep, and others (see Keay 1981). Other institutions, such as the Royal Asiatic Society, contributed much to an understanding of the richness of Asian culture and civilization. While there is much that is noteworthy in this scholarship, it is probably the case as well that it may have contributed to the emergence of the "Orientalist" mindset.

It can be argued that the very notion of a social science is a Western construct, arising in the late eighteenth century to describe and explain the far-reaching changes in European societies as a result of industrialization. Quantification of data was seen as an aid to explanation and rational decision making. Though there was, of course, much earlier writing on education issues and topics, a "science of education" to be used deliberately to address systemic education issues can be said to have developed about five decades ago as a result of two distinct forces. One was that decolonization raised the question of how viable postcolonial societies were to be built, and what contribution an educated citizenry and labour force could make to societal, political, and economic transformation. The other was the shift of teacher preparation to universities, which led in turn to the emergence of the applied disciplines, educational philosophy, educational psychology, sociology of education, and so on. Two major theories dominate this literature on the role of education in societal transformation. Modernization theory (Black 1966) took, we now see, an uncritical view of the contribution that the school system could make to modernization, failing to recognize the power of various other forces such as race, class, and gender to limit opportunity and mobility. Neo-Marxist analysis, however, saw schooling in capitalist societies as inherently inequitable, exploitative, and disempowering. Center-periphery theories (Frank 1978) saw education in the ex-colonies, the periphery, as nonagentic and incapable of transformation. The inadequacies of these approaches became apparent in the late 1980s when the economic transformation of Japan, followed rapidly by South Korea, Taiwan, Hong Kong, and Singapore, led to an examination of the education and training structures that underpinned the economic miracle (World Bank 1993).

Yet another perspective had emerged earlier in the 1970s, which was grounded in the center-periphery thesis. Here the argument was that the world system was inherently unequal, divided into centers—Western industrialized nations mostly with colonial pasts, and nation states mostly postcolonial, whose economies and politics were peripheral to the ways in which international economies and politics worked. A cultural-educational consequence of such inequality was that key educational and cultural resources from the West dominated and overwhelmed societies on the periphery resulting in a "colonization of the mind." Philip Altbach (1977), Ali Mazrui (1975), Syed H. Alatas (1977), Carnoy (1974), and Goonatilake (1982) have all detailed the ways in which hegemonic intellectual processes and metaphors have shaped education and intellectual practice in postcolonial societies. While offering a refreshing critique of dominant discourses on the reasons for Eastern underdevelopment, it does draw up a picture of nonagency and compliant behavior.

A third perspective looks at disciplines from a local knowledge perspective. Alan Bishop (1990), among others, has suggested that mathematics, seen as being culture-free and having universal validity, is actually "humanly constructed [and thus] has a cultural history" (72). Bishop points out that there are mathematical ideas and systems that have developed differently within different cultures, and the reason why Western mathematics triumphed was due to "trade, administration and education." He points out that a value system characterized by rationality and objectivism was closely tied to the colonial imperative to control and dominate. So successful was this effort as represented in the industrial mode of production that these values became the hallmark of modernity and led in turn to a demand for more Western-styled education.

It was always the case during the colonial period that there were local, nation-level processes of resistance, accommodation, and adaptation. The resistances to imperial authority led to these traditional ways of knowing and doing being condemned as nonrational and superstitious. What is new is that globalization, new communication technologies, the environmental crisis, among others, have created, almost paradoxically, new opportunities and avenues for indigenization, of the valuing of culturally appropriate knowledge, and of ways of knowing and understanding beyond the rational and the logical. Knowledge indigenization (Semali 1999 2002) is a complex set of processes, including ways of knowing (of comprehending reality and acting upon it) and ways of being, that is prevalent in a community. In the context of education, it is that which students bring into classrooms, and which is fully present in lives outside of classrooms. Indigenous knowledge, it must be noted, does not consist solely of traditional knowledge, but also encompasses ways in which nonindigenous knowledge has been adapted and domesticated to serve local national purposes. According to Canagarajah local knowledge is best thought of as fluid and relational, and not a unitary or homogenous construct. He argues that it is not always progressive or radical and needs to be critiqued "through an ongoing process of critical reinterpretation, counter-discursive negotiation and imaginative application" (Canagarajah 2002, 251), a useful reminder of the dangers of reifying traditional knowledge.

Knowledge indigenization as a concept is not new. Scholars working within the dependency and center-periphery frameworks (e.g., Altbach 1977) raised issues of inequalities in knowledge production and use almost three decades ago, often in the context of an analysis of the persistence of neocolonial processes. My own doctoral-level research work in the early 1980s was on how the disciplines of literature, sociology, and botany positioned themselves and adapted to different socioeconomic contexts in two developing nations (Gopinathan 1984). More recent work has come from scholars examining the effects of colonial languages such as English on schooling effectiveness and power relations, and others seeking explanations for deep differences in academic achievement. Scholars have debated how the persistence in the use of colonial languages, especially English, drowns out the voices of indigenous speakers (Pennycook 1994; Canagarajah 1999). Cortazzi (1998) points to "cultures of learning" that necessitate studying learners' behavior within their own cultural settings. Wang and Lin (2005) and Fan et al. (2004) argue that mathematics learning is a culturally scripted activity, by which they mean that socially held views about knowledge, both general and domain-specific learning, all of which are culturally embedded, influence teaching and learning approaches. Other researchers note that certain culturally based variables, such as the view that effort rather than

ability is the key to academic achievement, can be powerful drivers for
effective learning; this literature seeks to provide an explanation for differences in mathematics achievement between East Asian and Western-
schooled students.

There is another reason why the present time is conducive for the
argument in favor of knowledge indigenization in education. Much of
the recent educational literature related to the enhancement of learning
outcomes recognizes the need for learning to be meaningful for it to be
powerful and transformative. This is in part a response, both to a better understanding of learning processes bequeathed by cognitive science,
and of the recognition that an older subject-centered curriculum and
teacher-dominated pedagogy will not provide students with the learning
skills and opportunities that are needed to meet the challenges of twenty-
first-century living. Much greater attention is now being paid to the need
for students to critique and question codified knowledge as represented
in textbooks, to adopt a problem-solving stance, and for students and
teachers to co-construct knowledge. This opens up space to discuss what
forms of knowledge, as well as means for attaining and legitimating it are
useful and appropriate in educational contexts.

There is no doubting the fact that compared to several decades
ago, there is much more of a national character in Eastern education systems, even as significant regional disparities both within and
between countries exist; the postcolonial phase of rebuilding education systems is over in most countries. All school systems in the region
have a national curriculum, teacher training, examinations, and so on.
Everybody speaks of preparing students for the knowledge-based economy, for students to become critical and creative thinkers, for teaching
to be learner-centered, for assessment to be formative and authentic, for
more Information and Computer Technology (ICT) to be used, and
so on. Systems now have to cope with new challenges and each passing month sees a new reform initiative. Asian educational experts borrowing from Western literature speak of constructivism as the key to
enhanced learning, while teacher educators speak of the need for action
research and reflective practice. This is a major challenge in countries
such as Japan, South Korea, and Singapore where an efficient content-
and teacher-dominant model has produced social mobility, boosted
industrial prowess, and yielded high academic achievement. School
systems in these countries now have to make major changes to meet
economic and cultural globalization challenges. The key question that
is not being examined closely enough is what these concepts mean in
practice and their appropriateness to vastly different cultural-pedagogic
contexts.

A significant volume of studies entitled *School Leadership and
Administration: Adopting a Cultural Perspective* edited by Walker and

Dimmock (2002) makes a strong case for cultural authenticity. The editors boldly state that

> education leadership and management has developed along ethnocentric lines being heavily dominated by Anglo American paradigms and theories...A narrow ethnicity pervades research and policy...theory and policy in educational administration and leadership are possibly more strongly contextually bound than many...are prepared to acknowledge. (2)

In a singularly insightful chapter in the above book, David Watkins, offers a review of, and arguments for, strong attention to be paid to culturally rooted perspectives. Watkins asserts boldly:

> It is now recognized that many of the findings presented in psychology textbooks around the world are based on research with American psychology undergraduates, typically 20-year-old white males (Lonner, 1989). Moreover...Kohlberg's theory of moral development, Piaget's theory of cognitive development, Maslow's theory of self-actualization, and Hertzberg's theory of job satisfaction are based on the values of Western culture: in particular an individualistic, independent conception of the person. In one of the most influential papers to date in all the psychological literature, Markus and Kitayama (1991) argued that an interdependent conception of the self was more appropriate for most persons from non-Western cultures and this has implications for basic psychological processes such as cognition, motivation, and emotion. (61–62)

One may, therefore, legitimately ask to what extent and in what ways these commonly used constructs in educational psychology textbooks, which in turn form the basis for much advice on instructional strategies, have multicultural validity.

A fundamental problem that has not been faced up to by Asian educators and which may be contributing to ineffectiveness with regard to school change and teacher-education reform generally is the strong role of teachers' prior beliefs and how culture may be relevant here. Settled cultures, even when faced with modernization influences, do not change quickly or much with respect to values such as child-adult relationships, authority, obedience, and respect. If the strength of these traditional cultural features is not recognized and a Western-oriented pedagogic model is adopted, the lack of fit could have serious consequences. This is not to say that nothing changes; indeed cultural dissonance may provoke reflection and change but, if not explicitly recognized and used to foster reflection, it can result in confusion, and ultimately lack of acceptance. Hallinger and Kantamara (2002) provide evidence of change in Thai principals' leadership styles; principals who recognized the strength of cultural norms, for example, respect for authority, sometimes used them to good effect to bring out significant change in their schools.

I will look at just one set of research studies, in mathematics, to indicate how complex the issues of culture, pedagogy, and transfer are. The Third International Mathematics and Science Studies (TIMSS) reported that East Asian students outperformed students in the United States and other Western countries. The success of East Asian students led to a spate of studies in the 1990s to explain the phenomenon (Leung 1995; Serpell and Hatano 1997; Stevenson and Stigler 1992; Stigler and Fernandez 1995; Fan et al. 2004; Jones 2008).

These studies are beginning to provide rich and nuanced details on how differences in mathematics achievement can be understood. They move beyond stereotypical notions of rote learning and memorization as characteristic features of Eastern classrooms. They indicate that an interlinked set of variables, such as models of teacher preparation, teacher beliefs, pedagogic practices, cultural beliefs about student ability, value of effort, individual differences, and so on, needs to be understood in context to explain achievement differences. Lee's study (1998) looked at frequency of verbal explanation and evaluation by students, how teachers went about creating more engaged learning on the part of students— she considered teachers' use of a variety of examples, how they extended student answers and involved other students, how they related concrete operations to abstract concepts, and how they went about facilitating deeper understanding. Lee noted that "East Asian teachers also more frequently involved students in the evaluation of the relevance or correctness of their own or their fellow classmates responses" (54), and more generally that "East Asian teachers have more success applying the constructive approaches in their classrooms while fewer US teachers do" (64).

Watkins (2002) points to the work of Jin and Cortazzi (1998, 756) where they indicate that Western scholars' concerns that creativity is stifled in Asian classrooms is based on a misunderstanding. Chinese teachers, they argue, see creativity as a slow process dependent upon content mastery. The development of children's understanding is seen as a slow, not rapid, developmental process. In a similar fashion we need to carefully understand the nature of teacher talk, student-teacher interactions, and lesson and assessment routines as played out in Asian classrooms.

This rich literature on the influence of culture on teacher beliefs and fine-grained analysis of pedagogy that is now emerging has important implications for education and teacher preparation in the East and needs to be seriously considered. It is clear that pedagogic practice is still rooted in distinctive cultures and each country will have to invest in research to find out what works best in its contexts. For too long policy prescriptions have looked to distal variables—the role of textbooks, class size, and instructional time—and not enough at the ways in which teachers teach, instructional practice that impacts both on the quality of instruction, and on how much is learnt. There is not much point in reducing class size or increasing instructional time, if more of the ineffective pedagogy is dished out. While an analysis of theory-practice relationships, a central consideration in reforming teacher education, is important, more important perhaps is what goes into the theory and the practice.

Scholars and policymakers need to consider the adequacy of a Western model to the educational challenges faced in the East. A substantial change in pedagogic effectiveness has not occurred in several decades after the many curricular and pedagogic reforms initiated in many school systems. The central problematic in the reform of schooling and schooling practices in Eastern educational contexts is that we have failed to consider what would be appropriate, culturally relevant pedagogy for our classrooms, and how this pedagogy can be constructed. It would not be an unfair characterization to say that the recommended pedagogy is a curious mixture of traditional, tactically understood notions of what it means to teach and learn subject content interwoven with ideas and concepts imported from abroad. As the research cited earlier indicates, there is an emergent body of work that questions the appropriateness of some imported models, and how certain culturally grounded features of the local, particularistic environment are the key to fashioning appropriate pedagogies (Fang and Gopinathan 2009).

It is clear that teacher education, in particular, needs to incorporate these perspectives and literature more centrally in the preparation of teachers in the region. More boldly, Leung (2003, 2002) asks what an East Asian model of teacher education would look like; he suggests that the idea of the "scholar-teacher," which is deeply rooted in Chinese culture, might explain the emphasis on content mastery that is characteristic of teacher-preparation programs in China. Leung also suggests that this might be the reason why Chinese teachers largely see mathematics learning as mastery of mathematical content rather than process (cited in Mok and Morris 2001). More importantly, we need to explore what might be key constituent elements of an Asian model of pedagogy that would better reflect the sociocultural elements of distinct cultural regions.

This is a major task for curriculum planners and teacher educators, for there is little research to go on. We need to know, in each country/ regional context, what pedagogical practices will optimize student outcomes. We do not know very well (in a well-documented, empirical sense) what goes on in our classrooms, at the different levels, in the different disciplines, and in different regions in our countries. We do not know how effective the teaching strategies that we recommend are in the context of the learning challenges faced by pupils in diverse socioeducational environments. These learning challenges are not unitary in that mastery of essential content, and concepts may require at one level a certain set of strategies and, perhaps at higher levels, a different set of strategies for analyzing and critiquing knowledge and to foster attempts at applying knowledge to the solving of problems. All students need to be enabled to do complex cognitive tasks; what is at issue here is how, within specific socioeconomic pedagogical environments, teachers can facilitate this. Such research is particularly important at a time when many calls for reform seem to trivialize the traditional and privilege the new. Teacher trainees, therefore, need to see instructional strategies not as ready-made answers to the challenges of learning but more tentative and provisional; taking such a position will make them more prepared to be flexible, more ready to adjust to the circumstances they find themselves in, and less prone to blame their students for the failure of their "best methods." How well trainees internalize this knowledge is both a matter of content and of the pedagogy utilized by teacher trainers.

One major difficulty that we will face is that we have invested too little in the types of educational research that will make this kind of teacher educational reform possible. We do not have sufficiently comprehensive knowledge about say, classroom pedagogy, about the impact of nonschool factors on learning, about teacher "folk pedagogic beliefs," and about the impact and persistence of certain types of training, to make the changes possible. Yet at the policy level recommendations for change come thick and fast—integrated curriculum, school devolution, school-based curriculum development, active learning, group work, authentic assessments—the list goes on and on! There is often a naive belief in the universal applicability of these constructs across vastly different cultural and epistemological contexts. Belief in the appropriateness of these constructs is not based on comprehensive and rigorous research within national socioeducational contexts. For education reform to be truly successful within a distinctive national context that is attempting major changes in learning environments, we need appropriate teacher education, and that in turn needs a strong body of research-based evidence that would shed light on what would constitute effective pedagogy and common standards of practice within teacher education.

How would a dialogue of cultures/civilizations approach benefit education? At the risk of oversimplifying a vast literature and complex processes, it should be pointed out that a major characteristic of Eastern "theorizing" on the value, purposes, and processes of education is that education is primarily about the cultivation of character, of emancipation of self and community.

Tagore's ideas on education illustrate this well. He did not deny the value of science and its relevance to raising the wealth and quality of life in India's villages, but he insisted that it was equally important to provide for a creative and cultural life. Elmhirst (1961) writes of Tagore that he believed that children needed to be stirred by an education built around their own need to grow, to imagine and to explore. The indirect effect of their activity upon the age-long customs and attitudes of their parents might be revolutionary. The parents might be won from a negative defeatism and dependency to a positive pride in the achievement of their children...It was Tagore who encouraged more and more survey and investigation, not for its own sake by theorists, but directed all the time towards specific problems.

These perspectives would be a powerful antidote to the prevailing orientations in modernist West-inspired education systems. In most postcolonial societies, driven by the urge to modernize, a utilitarian mentality pervades education. The mantra is that education is vital for economic growth and competitiveness; it is rates of return on investment that justify levels of spending on different types of education and training, and because education is treated as a positional good, parents in better-resourced social classes invest heavily in education, and often have access to better-quality private schools. Schooling has become excessively competitive, especially in East Asian societies; as a result, great effort is required from and great pressure is imposed on students. Often, high performance is not matched by interest or engagement. Curiosity, imagination, and innovation having been squeezed out by the pressure to master content is now being reintroduced via techniques such as mind maps and creativity workshops!

The argument above is not about replacing a Western model with an Eastern one; it should be seen as a call to build upon the strengths of both. Many Western theorists of education would wholeheartedly agree with Tagore's views. A commitment to reason and evidence and an emphasis on science and technology, among others, have contributed to Asia's march to modernity. Twenty-first-century global realities are that we live in an interconnected and interdependent world. There are many areas in the Eastern developing world that would benefit from greater access to a reason-based Western model. Traditional models such as madrassas need a dose of reform. Equally, in Western systems there is need for a greater knowledge and understanding of non-Western cultures if they are to productively come to terms with an emergent and revitalized East.

In conclusion, then, it is clear that there is a need to rethink education frameworks in the context of a complex, interdependent world. The circumstances are right for such a task, principally because there continues to be severe failings in current education models, and also because globalization and changes in technology have made the local more visible. More importantly, the emergence of critiques of modernization from dependency theorists, postcolonial scholars, and feminists has opened up space to rethink education more radically, and to push the indigenization idea more forcefully. It is argued that teaching is a culturally constructed activity, and for teaching to be more empowering it needs to be more culturally authentic.

Bibliography

Alatas, Syed Hussein. 1977. *Intellectuals in Developing Societies.* London: Frank Cass.

Altbach, Philip G. 1977. "Servitude of the Mind? Education Dependency and Neo-Colonialism." *Teachers College Record* 79: 187–204.

Bishop, Alan J. 1990. "Western Mathematics: The Secret Weapon of Cultural Imperialism." *Race and Class* 32: 51–65.

Black, C. E. 1966. *The Dynamics of Modernization: A Study in Comparative History.* New York: Harper and Row.

Canagarajah S. 2002. "Reconstructing Local Knowledge." *Journal of Language, Identity and Education* 1: 243–259.

———. 1999. *Resisting Linguistic Imperialism in English Teaching.* Oxford: Oxford University Press.

Carnoy, Martin. 1974. *Education as Cultural Imperialism.* New York: Longman.

Chen, Sharon Hsiao Lan. 2001. "Constructing a Constructivist Teacher Education: A Taiwan Experience." In *New Teacher Education for the Future: International Perspectives,* edited by Yin Cheong Cheng, King Wai Chow, and Kwok Tung Tsui, pp. 261–290. Hong Kong: Hong Kong Institute of Education/Kluwer Academic Publishers.

Cheng, Yin Cheong, King Wai Chow, and Magdalena Mo Ching Mok. eds. 2004. *Reform of Teacher Education in the Asia-Pacific in the New Millennium: Trends and Challenges.* Dordrecht: Kluwer Academic Publishers for the Hong Kong Institute of Education and the Asia Pacific Educational Research Association.

Cortazzi, Martin. 1998. "Learning from Asian Lessons: Cultural Expectations and Classroom Talk." *Education* 26: 3–13.

Elmhirst, L. K. 1961. *Rabindranath Tagore, Pioneer in Education: Essays and Exchanges between Rabindranath Tagore and Elmhirst.* London: John Murray.

Fan, Lianghuo, Ngai-Ying Wong, Jinfa Cai, and Shiqui Li. eds. *How Chinese Learn Mathematics: Perspectives from Insiders.* Singapore: World Scientific, 2004.

Fang, Yanping, and S. Gopinathan. "Teachers and Teaching in Eastern and Western Schools: A Critical Review of Cross-Cultural Comparative Studies." In *International Handbook of Research on Teachers and Teaching,* edited by Lawrence J. Saha and A. Gary Dworkin, pp. 557–572. New York: Springer, 2009.

Frank, A. G. 1978. *Dependent Accumulation and Underdevelopment.* London: Macmillan.

Gopinathan, S. 1984. "Intellectual Dependency and the Indigenization Response: Case Studies of Three Disciplines in Two Third World Universities." PhD Dissertation. Buffalo: State University of New York.

Goonatilake, S. 1982. *Crippled Minds: An Exploration into Colonial Culture.* New Delhi: Vikas Publishing House.

Hallinger, Philip, and Pornkasem Kantamara. 2002. "Educational Change in Thailand: Opening a Window onto Leadership as a Cultural Process." In *School Leadership and Administration: Adopting a Cultural Perspective*, edited by Allan Walker and Clive Dimmock, pp. 123–140. London: RoutledgeFalmer.

Hatano, Giyoo, and Kayoko Inagaki. 1998. "Cultural Contexts of Schooling Revisited: A Review of The Learning Gap from a Cultural Psychology Perspective." In *Global Prospects for Education: Development, Culture, and Schooling*, edited by Scott G. Paris and Henry M. Wellman, pp. 79–104. Washington, DC: American Psychological Association.

Jin, Li, and Martin Cortazzi. 1998. "Dimensions of Dialogue: Large Classes in China." *International Journal of Educational Research*, 29: 739–761.

Jones, Keith. 2008. "Windows on Mathematics Education Research in Mainland China: A Thematic Review." *Research in Mathematics Education* 10: 107–113.

Keay, John. 1981. *India Discovered: The Recovery of a Lost Civilization.* London: Harper Collins.

Lee, Shin-Ying. 1998. "Mathematics Learning and Teaching in the School Context: Reflections from Cross-Cultural Comparisons." In *Global Prospects for Education: Development, Culture, and Schooling*, edited by Scott G. Paris and Henry M. Wellman, pp. 45–77. Washington DC: American Psychological Association.

Leung, Frederick K. S. 2003. "Issues Concerning Teacher Education in the East Asian Region." *Asia-Pacific Journal of Teacher Education and Development* 6, no. 2: 5–21.

———. 1995. "The Mathematics Classroom in Beijing, Hong Kong and London." *Educational Studies in Mathematics* 29: 297–325.

———. 2002. "In Search of an East Asian Identity in Mathematics Education." *Educational Studies in Mathematics* 47: 35–200.

Mazrui, Ali. 1975. "The African University as a Multinational Corporation: Problems of Penetration and Dependency." *Harvard Educational Review* 45: 191–210.

Ming-Tak, Hue. 2005. "The Influences of Chinese Culture on Teacher-Student Interaction in the Classrooms of Hong Kong Secondary Schools." *Curriculum Perspectives* 25: 37–43.

Mok, I. A. C., and Morris, P. 2001. "The Metamorphosis of the Virtuoso: Pedagogic Patterns in Hong Kong Primary Mathematics Classrooms." *Teaching and Teacher Education* 17, no. 4: 455–468.

O'Sullivan, Margo. 2004. "The Reconceptualisation of Learner-Centred Approaches: A Namibian Case Study." *International Journal of Educational Development* 24: 585–602.

Pennycook, Alastair. 1994. *The Cultural Politics of English as an International Language.* London: Longman.

Richardson, Patricia M. 2004. "Possible Influences of Arabic-Islamic Culture on the Reflective Practices Proposed for an Education Degree at the Higher Colleges of Technology in the United Arab Emirates." *International Journal of Educational Development* 24: 429–436.

Semali, Ladislaus M. 2002. "Cultural Perspectives and Teacher Education: Indigenous Pedagogies in an African Context." In *Teacher Education: Dilemmas and Prospects, World Yearbook of Education*, edited by Elwyn Thomas, pp. 155–165. London: Kogan Page.

———. 1999. "Community as Classroom: Dilemmas of Valuing African Indigenous Literacy in Education." *International Review of Education* 45: 305–319.

Serpell, R., and G. Hatano. 1997. "Education, Schooling, and Literacy." In *Handbook of Cross-Cultural Psychology* vol. 2, edited by John W. Berry, Pierre R. Dasen, and T. S. Saraswathi, pp. 339–376. Boston: Allyn & Bacon.

Stevenson, Harold W., and James W. Stigler. 1992. *The Learning Gap: Why Our Schools Are Failing and What We Can Learn from Japanese and Chinese Education*. New York: Summit Books.

———, James W. Stigler, Shin-ying Lee, G. William Lucker, Seiro Kitamura, and Chen-chin Hsu. 1985. "Cognitive Performance and Achievement of Japanese, Chinese, and American Children." *Child Development* 56 : 718–734.

Stigler, James W., and Clea Fernandez. 1995. "Learning Mathematics from Classroom Instruction: Cross-cultural and Experimental Perspectives." In *Basic and Applied Perspectives on Learning, Cognition, and Development. The Minnesota Symposia on Child Psychology* vol. 28, edited by Charles A. Nelson, pp. 103–130. Hillsdale, NJ: Lawrence Erlbaum Associates.

Walker, Allan, and Clive Dimmock. eds. 2002. *School Leadership and Administration: Adopting a Cultural Perspective*. London: RoutledgeFalmer.

Wang, Jian, and Emily Lin. 2005. "Comparative Studies on US and Chinese Mathematics Learning and Implications for Standards-Based Teaching Reform." *Educational Researcher* 34, no. 5: 3–13.

Watkins, David. 2002. "Learning and Teaching: A Cross-Cultural Perspective." In *School Leadership and Administration: Adopting a Cultural Perspective*, edited by Allan Walker and Clive Dimmock, pp. 61–76. London: RoutledgeFalmer, 2002.

World Bank. 1993. *The East Asian Miracle: Economic Growth and Public Policy*. Washington, DC: Oxford University Press.

14

SOUTHEAST ASIA'S INDIGENOUS KNOWLEDGE: THE CONQUEST OF THE MENTAL TERRA INCOGNITAE

Victor R. Savage

INTRODUCTION

The crucial feature of human life is its fundamentally *dialogical* character. We became full human agents, capable of understanding ourselves, and hence of defining our identity, through our acquisition of rich human languages of expression—where the latter include, in addition to spoken and written language, also the "languages" of art, of gesture, of love, and the like. (Fred Dallmayr 2002, 60)

Fred Dallmayr's quotation captures the long dialogue between global communities that were either consciously or unconsciously creating a growing repository of human knowledge not only about the world around them, but also human-nature relationships—in short, "science" in its various incarnations. The forms of dialogue, Dallmayr suggests, involve not only exchanging knowledge through writing, scientific symbols, and figures, but also include a whole range of expressive human communication transcending the written language. After all, for most of human history *homo sapiens* survived as an illiterate species, but that did not stop human communities from exchanging information, knowledge, perceptions, goods, and experiences about their ecosystems, nature, and the cosmos.[1]

Today the indigenous knowledge embodied in preliterate cultures—especially their environmental knowledge through which they define the relationship between society and nature, culture and ecology—have become central concerns within academic discourse after environmental degradation and climate change have come to be perceived as major global challenges. This chapter will explore Southeast Asian indigenous knowledge and how it can be integrated into environmental education in the region.[2]

The Southeast Asian region, which has 3 percent of the world's land area and 7.7 percent of the global population, is at the confluence of the indigenous traditions of Chinese, Indian, Middle Eastern, and European cultural and civilizational inputs. It is also at the crossroads of many of the world religions: Buddhism, Hinduism, Islam, Judaism, and Christianity as well as Chinese religions. Despite being illiterate, indigenous groups in the region have captured significant natural and cultural events of their ancestry in their myths and legends. For example, the Ilongot head-hunters in Luzon (Philippines) have myths and folk stories reflecting the impact of the Second World War upon their cultures (Rosaldo 1994). To take another example, there are over a hundred tribal myths, including those of Taiwan Aboriginals, Batak, Iban, Flores, and Benua-Jakun, as well as in the Indian *Puranas* and the epic *Mahabharata*, which refer to a rise in prehistoric sea levels in the form of the narratives of a "great flood." Stephen Oppenheimer (1999, 263–296) has controversially argued that the events in these myths refer to an actual great flood that occurred when the ice caps melted during interglacial epochs, and that these floods are also the source of the biblical story of Noah and the great Ark. He also claims that communities inhabiting the fertile valley of the Sunda shelf adapted to the rising interglacial sea levels 14,000 years ago by migrating to higher lands in insular and mainland Southeast Asia (Oppenheimer 1999, 23–48). Thus indigenous communities despite being illiterate are *Earth literate,* and their myths carry important historical knowledge of environmental changes. Given our current interest in climate change, one can use indigenous folk documentation of sea level rises in the past, for example, to construct an understanding of its impacts in the region.

In this chapter, I would like to point out four major areas where indigenous knowledge in the region can make contributions to explorations of Nature and our relationships with ecosystems: (1) the cultural appraisals of natural resources; (2) the understanding of tropical environments and landscapes; (3) astronomy, celestial mapping, and navigation; and, (4) herbal medicines. I will also briefly consider how such knowledge could be incorporated into environmental education programs in the region.

THE CULTURAL APPRAISAL OF
NATURAL RESOURCES

The renowned American geographer, Carl Sauer (1969, 2–3) noted that "natural resources are cultural appraisals." He used this paradoxical expression to make a clear distinction between notions of nature and environment that are neutral qualifications, and the notion of natural resources that refers to humanly defined and valued elements of nature. For many indigenous communities the ability to turn environment into natural resources is a critical mechanism for survival.

Clearly, the diversity of flora and fauna, the multitudinous ecosystems, and varying physical landscapes of land and water in the region was a daunting challenge in the beginning for many communities in the region. While there are some 40,000 species of tropical plants identified in the world, 30,000 are found in Indonesia, of which 7,000 have become cultivated species. Over centuries of countless experiments with nature, Southeast Asian communities enlarged the repertoire of natural resources available for use in their diverse tropical environments. They learnt to utilize a vast array of flora and fauna for food, drink, beverage, liquor, cordage, color dyes, poisons, kitchen utensils, skin ointments, folk medicines, building materials, religious offerings, cloth, fish traps, and weapons. While one might be tempted to conclude that the diversity of nature created a diversity of natural resources along environmentally deterministic lines, I would suggest that the diversity of Southeast Asia's tropical landscapes offered only opportunities for human use—the onus was on Southeast Asian communities to experiment with this natural diversity to actualize its potential for providing natural resources.

Indeed the region has had a long phase of folk scientific experimentation that the American archaeologist, Wilhelm Solheim (1970), labeled the "lignic phase" or wooden and vegetative phase in the region's prehistory. This period of experimentation that expanded human understanding of nature's diverse potentials began some 22,000 years ago. It has continued till today. It is especially the people of the forest (hunters and gatherers), the people of the seas (sea gypsies or nomads), and the people of the mountains (swidden cultivators) who are the major repository of folk knowledge and folk science of the diverse tropical ecosystems in Southeast Asia. Many ethnographic studies extending over the last century have meticulously documented the cognition, knowledge, cultural appraisal, and utilization of tropical ecosystems by Southeast Asian communities (see Howell 1989; Tsing 1993; Conklin 1957; Dove 1990). It is through long experimentation to acquire knowledge of local flora and fauna that Southeast Asia's indigenous communities came to know about the specific qualities of the products in tropical nature, both on land and water.

The discovery of the unique "lignic" or vegetative tradition of the region has also had another consequence. It helped to alter earlier archaeological and prehistoric views of the region that tended to see Southeast Asia in a rather dismal light because, unlike other prehistoric traditions, the region's lithic (Neolithic and megalithic) or stone tradition seemed to have been stalled and stunted. Since the artifacts of lignic culture decay over time, unlike lithic artifacts that survive for millennia, it may seem little progress occurred during a lignic phase.

Subsequently the expansion of growing trade with foreigners transformed local produce into commodities of trade, monetary value, and capitalistic goods. Southeast Asia became a central trading port of call for many centuries before the advent of European colonialism. In precolonial times, the region was an important source of gold (see Wheatley, 1961), and by the sixteenth century Europeans turned the islands into the spice capital of the world. Colonialism created new economic values for local produce. Yet, it is not the famed gold or spices that capture the local widespread knowledge of plants and animals. It is the swidden or shifting cultivators, variously referred to as highlanders, hill tribes, and ethnic minorities, for example, the Thai *chao khao*, that really demonstrate indigenous knowledge of the widest diversity of plant types. Various studies of hill-tribal swidden communities have continually recorded the vast variety of crops they grow. Edward Anderson (1993, 56) recorded at least 88 different species of plants that swiddeners grow in northern Thailand. The vast range of crops grown provide such communities almost subsistence autonomy and sustainability, by providing for food, beverage, liquor, vegetables, cordage, fruits, cash crops, matting materials, housing materials, dyes, poisons, and folk medicines.

One of the most ubiquitous themes in the Chinese literature of the region lies in the economic assessment of various ports, kingdoms, and islands. For China, Southeast Asia through the centuries remained an important source of natural resources to satisfy her wide cultural and gastronomical needs for food, fruits, medicine, religion, aphrodisiac drugs, aesthetics, perfume, and spices. If India provided the civilizational catalyst for the region, China provided Southeast Asia the economic impetus and cultural appraisal of her diverse natural resources. This was the region of camphor, dragon's blood, benzoin, dammar, liquid storax, gardenia flowers, gharuwood, sandalwood, cloves, nutmegs, areca nuts, ebony, "sapan-wood," cardamoms, pepper, cubebs, aloes wood, coral-trees, pearls, ivory, rhinoceros horns, kingfisher feathers, ambergris, tortoise and turtle shells, and beeswax, among other products. In a quote from the *History of the Yuan Dynasty* (Book 210), one gets a glimpse of the value that the Chinese placed on the region's products when it mentions that "as a rule the barbarian countries over the sea produce many rare and valuable things, which fetch a high price in China" (quoted in Groeneveldt 1876, 246). It is also not surprising that one of the earliest scientific books on tropical and subtropical botany in the world was written in 304 CE by a Chinese official, Chi Han that was based substantially on tropical plants in Southeast Asia (Li 1979). To a large extent, the Chinese knowledge of the region's tropical nature came from its long colonization of Vietnam that it occupied for 1000 years (111 BCE–939 CE).

The Understanding of Tropical Environments and Landscapes

Scientific understanding of the region developed throughout the history of exploration in the region by foreign travelers, ship-captains, pilgrims, merchants, and colonial bureaucrats. Chinese, Indian, Arab, and finally European, sojourners all contributed to the history of such scientific observation: writing, mapping, drawing, sketching, and documenting their sensual representations. For Nayan Chanda (2007), the movements of varied peoples in search of conquests, trade, converts, and even personal ambition are stories of the globalization of the world, creating networks of traders, migrants, pilgrims, and consumers for the exchange of products, ideas, technology, and religious beliefs across borders. The Western explorations in the region since the sixteenth century enlarged the Western *oikoumene* and continued the process of globalization. The sixteenth-century Portuguese discovery of the Spice Islands, and their garrisons in Melaka and Timor, laid the foundations of early Western knowledge of this tropical region, and encouraged other European powers, such as the Spanish, Dutch, British, and French to similar explorations. So long as spices remained a prized commodity for the West, the island world remained a theater of attraction, exploration, discovery, and interaction for Western explorers, adventurers, naturalists, and sojourners, and in the process three developments took place.

First, the island world of Southeast Asia became an intense arena of Western territorial competition that led inevitably to rapid Western cartographic representation of this vast archipelago. Without the inputs of local mariners, traders, and seamen who had for centuries plied these seas, the Western cartographic representations would not have been so quickly and accurately reproduced by the end of the sixteenth century. Second, with the avid interest in spices, the island world's flora and fauna became a source of attention for various Western naturalists and administrators. We see scientific documentation of flora and fauna by the seventeenth and eighteenth centuries. With the help of the accumulated indigenous inputs and wisdom, the first comprehensive compendium of flora, fauna, and marine organisms was put together by the Blind Seer of Ambon, Georg Eberhard Rumphius or Rumpf (1628–1702), with probable help from his native wife and daughter. Despite being blind in 1670, Rumphius (1705) produced a major tome on plants and marine organisms in and around the island of Amboina entitled *Het Amboinsche kruidboek* or *Herbarium Amboinense* that documents 1,200 species of which 930 have definite species names and 140 were identified to genus level. The illustrations and classification of 350 plants were later included in the scientific classification of Linnaeus. Rumphius's work was almost copied in full and popularized by his student Francois Valentijn (1724–1726) in his five-volume *Oud en Nieuw Oost-Indien* (Old and New East Indies) that provided an encyclopedic listing of the flora, fauna, and marine organisms of Indonesia at that time.

The other important interest in the island world for Westerners was the beautiful bird of paradise, located in the Aru Islands and New Guinea. Once again, one sees cross-cultural comparative appraisals of nature extending far back in time. The bird of paradise was said to be the main product of specialist traders about 250 CE. Their feathers were used in the headdresses of indigenous dignitaries. By 1500 CE. Portuguese records note that bird-of-paradise feathers were sold in Persia and Turkey to decorate headdresses of important officials (Spyer 2000, 25). This native appraisal of the bird-of-paradise's feathers fed Western fascination for use in their hats (Savage 1984). This is a clear example that even the aesthetics and fashion of illiterate "primitives" of one culture can shape aesthetic taste of literate peoples in so-called more-advanced cultures. Patricia Spyer (2000, 63–64) has argued that the act of clothing native populations was one way in which Europeans demonstrated their colonial authority, their "civilizing mission," and a "tropical reflection of themselves." We could similarly argue that in the case of the use of bird-of-paradise feathers the reverse is true: the "primitives" were dressing their "civilized" counterparts, including their leaders and royalty. This Western fascination with the bird of paradise also resurrected Christian biblical views of paradise as located in the East, and made the bird of paradise the subject of much scientific curiosity. In the process, one of the earliest natural historians to document the birds in their native habitats in New Guinea was the French explorer Pierre Sonnerat (ca. 1748–1814). He also was one of the first ornithologists in the region to document in drawings and written descriptions various birds in New Guinea (Sonnerat 1776).

By the nineteenth century, the Western attraction in the region shifted from the island world to mainland Southeast Asia, or what is now called "Indochina." Once again, beside the competition for territory and natural resources, there was an added incentive of finding a back door into the lucrative Chinese Yunnan province. This search created major geographical expeditions and explorations in the region. The most famous of these explorations was the French Doudard de Lagrée-Francis Garnier Mekong expedition (1866–1868). While the expedition did not find a back door for Sino-French trade, it opened up a broad vista of scientific and cultural knowledge in the process. Of the 9,960 km covered by the expedition, 6,720 km were routes revealed for the first time, of which 5,060 km were discovered by Francis Garnier himself (Savage 1984, 61).

The expedition also collected some 1,500 to 2,000 new species belonging to the vegetable kingdom (Savage 1984, 107). While the expedition and its findings was an eye-opener for Western civilization, it certainly was not virgin territory for local communities. This was a landscape of many past indigenous civilizations and cosmic kingdoms, sedentary societies, and swidden communities. Given that it was virgin territory to Western scientific eyes, Southeast Asia in the nineteenth century remained a naturalist paradise, with numerous naturalists boasting their new finds. The legions of notable naturalists, botanists, and zoologists that contributed to the fund of scientific knowledge included Albert Bickmore, Henri Mouhot, William Hornaday, Carl Bock, Alfred Wallace, Eric Mjoberg, Fedor Jagor, Arthur Adams, John Anderson, Joseph Arnold, Robert Shelford, John Whitehead, Thomas Burbidge, Henry Forbes, Thomas Horsefield, Charles Hose, and Dean Worcester (see Savage 1984).

When the Industrial Revolution took off in the nineteenth century, the West became interested in tropical ecosystems for mining, forestry, and agriculture. Tropical science had become important for four reasons: finding ways for the white population to withstand tropical diseases through tropical medicine; creating adaptive environmental conditions to reduce environmental diseases; developing cash crops and raw materials through a scientific plantation system; and finding a more efficient and scientific means of logging wood. The most active Western colonial power to develop tropical agriculture was the Dutch. They introduced, developed, and expanded a whole host of cash crops in Indonesia—tea, coffee, sugar, cinchona, indigo, oil palm, rubber, tobacco, and pepper (see Geertz 1971, 38–82). The crowning achievement of the Dutch in tropical agriculture was the establishment of a tropical science research institute at the Botanical Gardens at Buitenzorg (now Bogor), known then as 's Lands Plantentuin or the Government's Botanical Gardens. The Botanical Gardens, was first established in 1817 for aesthetic, scientific, and educational reasons, and between 1884 and 1934 attracted 237 scientists representing 17 nationalities who contributed to expand understanding of not only the flora, but also the fauna and marine biology of the region (Savage 1984, 108). For the Western world, the European-styled botanical gardens that sprang up all over the region reflected in Denis Cosgrove's (2008, 53) opinion, "Europe's larger imperial project, embracing the history of colonization in much more extensive ways" than bringing back botanical curiosities, and commercial or medicinal plants from new worlds to Europe. Yet, the tropical botanical gardens were not mere emblems of colonized entities, because the variety of local plants in these gardens was a reflection of accumulated indigenous knowledge and relationships with tropical nature. The botanical gardens in my view can be seen as sites of the ongoing cultural dialogue between Europeans and native populations. Native dooryard gardens and swidden *ladangs* (cultivated fields) that had existed for centuries now became the subject of European scientific enquiry, landscape aesthetics, emblems of civilization, and symbols of colonization.

ASTRONOMY, CELESTIAL MAPPING, AND NAVIGATION

When the French intellectual Simon de la Loubere (1693) visited Siam (Thailand) in the 1680s, he found a high standard of astronomical science, linked to astrology and mathematics that astonished him. The Thai courts' understanding of astronomy would not surprise people familiar with the prehistory and history of the region. Southeast Asian communities as far back as 6,500 BCE were already plying the seas and oceans. These were the communities that invented the outrigger boats and finally formed settlements in Melanesia, Micronesia, Polynesia, and Madagascar and became the original inhabitants of these islands. But the main software of these tropical Vikings was not their boat technology but their folk astrological representations that served as their cartographic bearing for their long-distance journeys. Their interest in cosmologies is reflected in many applied ways in the region such as the development of astronomy and astrology, fortune-telling, celestial mapping, navigational aids, geomancy or Chinese feng shui (wind and water), calendars, cosmic-based architecture, and cosmic-based urban planning.

Prehistoric communities of the region had strong cosmic worldviews of a tripartite vertical space (upper world, human world, and lower world) in contrast to modern ideas of horizontal and territorial space (Savage 2009). Their interest in heavenly bodies continues today amongst hunting, gathering, and swidden agricultural tribal groups who deal with many sky, thunder, lightning, sun, moon, and rainbow gods, spirits, ancestors, and deities. The star designs on the bronze *dongson* drums found throughout the region also underscore the importance of the celestial sphere in Southeast Asian–prehistoric worldviews.

These indigenous prehistoric folk sciences of astrology and astronomy were over the centuries given a great boost with inputs from Chinese, Indian, and Arabic astronomical ideas. The Chinese interest in the geomantic science of feng shui, their astrological interests in animal horoscopes, and their *yin-yang* perspectives have all influenced the region. The Chinese astronomical knowledge had its most profound impact on Vietnam, given that the country was one of China's southern provinces for 1,000 years. There are 27 historical documents in Vietnam dealing with astronomy. Most of the documents reflect a combination of Chinese source materials and folk experiences of the Vietnamese (Nguyen 2002, 550). The Vietnamese also borrowed a lot of the Chinese ideas on geomancy, which is widely practiced in East Asia for locating and building ancestral graves, houses, palaces, temples, and cities (Yoon 2006). In Vietnam, there are 84 historical treatises on geomancy, whose authors come from both China and Vietnam (Nguyen 2002, 552). Many hill-tribe communities in the region have also adopted Chinese feng shui (wind and water) locational practices in their migratory and temporary settlements as they engage in swidden agriculture. For example, the Akha had the practice of dropping an egg in a cleared area to see if their ancestral spirits and the local "Lords of Land and Water" approve of their location for settlement (Anderson 1993, 149).

Given the region's long prehistoric interest in cosmic elements (stars, solar and lunar systems, and weather elements), the region has developed its own indigenous calendar system. The use of traditional calendars is found in Burma, Thailand, Laos, Cambodia, Vietnam, and Indonesia. The Southeast Asian calendar is based on a 19-year cycle, with each month consisting of 30-days (exactly 29.5 days), and one year consisting of 365.3 days (Yukio 2002, 397–399). Ôhashi Yukio (2002) hypothesizes that the 19-year cycle of the Southeast Asian calendar is a hybrid of both Chinese and Indian influences. He argues that the Southeast Asians accepted both Chinese and Indian astronomy, and that the region's 19-year cycle was developed probably from inputs of the Chinese Chu (Sipong-panna method) calendar and the Indian calendar (Ardharatrika school) (Yukio 2002, 401–402). However, what is still missing in research is the history of Southeast Asian astronomy and how indigenous knowledge informed the regional calendar that developed.

SOUTHEAST ASIAN FOLK MEDICINES: THE MULTICULTURAL DISCOURSE

One of the areas of cultural and scientific exchange between indigenous communities and their counterparts in the wider Asian region is "folk medicine." Unlike Western medicine, Southeast Asian folk medicines were holistic systems of body care, cures against illnesses, longevity preservation, and beauty treatments. Hence, the Indonesian traditional medicinal practice of *jamu* underscores this holistic and comprehensive system of inner and outer health and beauty encompassing powders, pills, ointments, lotions, massage, and ancient folklore (Beers 2001, 7).

Folk medicinal practices in the region identify three causes of disease: natural causes (stabbing, broken bones, and food poisoning); supernatural causes (spiritual origins of disease), and metaphysical causes. While all three causes have indigenous origins, the metaphysical cause reflects the influence of Chinese medicine. In many hill-tribal and peasant communities, the Chinese Taoist concept of *yin-yang*, and the idea of the balance of hot and cold in foods and drinks, correlated with age and gender, are commonly seen as informing practice in herbal medicines.

Hence, "hot" diseases are treated with "cold" medicines, and "cold" ailments are cured with "hot" medicines (Anderson 1993, 129). For the Lisu, cooling tea is boiled for a patient who is "hot" inside (Anderson 1993, 129). Similarly, the Indonesian *Jamu* is a holistic therapy based on the Chinese concept of "harmony." Specifically therapies are aimed at finding a balance between persons and their environments and between hot and cold elements of the body, and medicinal herbs are divided into hot and cold categories (Beers 2001, 29). The close association between China and Vietnam over the centuries has led to much exchange in the knowledge and practice of traditional medicines between the two regions. The Vietnamese have major catalogs of traditional medicines informed by principles of Eastern medicines, with over eight hundred specimens of local and northern herbal medicines, as well as other prescriptions including popular medicinal wisdom (Lan 2002, 266–267). The Vietnamese have also contributed to the discovery of 15 new points included in Chinese acupuncture (Lan 2002, 266).

An important characteristic of Chinese medicine is its holistic and organic view of diseases and cures. At the center of the East Asian perspective of nature, cosmos and folk medicine are the notions of *Qi*, the mixing of the spirit of the *yin* (earth) and *yang* (heaven) forces that cause all changes through the gathering and dispersing of *Qi*, thereby producing all things in the Universe (Lee 2002, 61). Given that *Qi* is the life force in all nature, including human beings, it is seen as the source of human life, energy, physical stamina, as well as the cause of illness and diseases by its excess or deficit. *Qi* might be equated with the Hindu universal goddess *Sakthi*, an energy force that pervades all life in the universe. In Southeast Asia, various communities and tribal groups have different variants of belief in such a universal life force. For example, the Chewong of Peninsular Malaya believe in *ruwai*—a "vital principle" found in all human beings including the fetus (Howell 1989, 55).

Given their long association with plants in the region, tribal groups in mainland Southeast Asia have a huge pharmacopeia of nearly seven hundred medicinal plants for treating all sorts of illnesses, diseases, and pains (Edward Anderson: 1993, 127). In Indonesia, close to 1,000 plants are used in traditional medicine (Beers 2001, 57). Over the centuries, tribal groups in mainland Southeast Asia have discovered a wide variety of plant-derived drugs to combat a variety of diseases: fevers, pains, bites and stings, skin problems, burns, nausea, nosebleed, earache, dizziness, paralysis, and even hypertension (Anderson 1993, 143). In Java, besides treating all sorts of illnesses and diseases, herbal medicines are used for cosmetic purposes including the promotion of soft and glowing skin care, weight reduction, coping with hangovers, tightening the vagina, enhancing the male libido, nurturing sexual energy, and increasing the size of penis erection (Beers 2001, 30–32). Indeed, long before Viagra was discovered in 1998, Southeast Asian men have been using all sorts of drugs and aphrodisiacs to enhance their sexual potency and energy. For Indonesian women, staying beautiful, healthy, and sexually potent is all part of their belief in maintaining large families to ensure their economic and social prestige in their communities.

Javanese herbal medicine is as old as Javanese civilization and originated in the ancient palaces of Surakarta (Solo) and Yogjakarta. These traditional folk medical traditions in Central Java are the product of indigenous developments mingling with Chinese, Indian, and Arabic influences (Beers 2001, 13). The development of folk medicine has been documented in several indigenous treatises on the subject. One of the most comprehensive Javanese medicinal texts is the *Serat Kawruh bab Jampi-jampi* (A Treatise on All Manner of Cures) that provides 1,734 formulae of *jamu* made from natural ingredients (Beers 2001, 15). In 1977, a research team in Kendari, Southeast Sulawesi found some 449 herbal remedies used (Beers 2001, 50). In Ceram, 30 species of medicinal plants were commonly used. On the Borobudur monument constructed c. 800–900 CE, we can see bas-relief carvings of the realm of "form and desire" (our world) depicting the *kalpataruh* leaf, from the mythological tree that never dies, together with other ingredients for women's health and beauty. Belief in herbal medicines continues to be strong in many parts of the region despite the advance of Western medicines, and no traveler or tourist would be able to resist the many traditional massages found all over the region: in Thailand, Laos, Cambodia, Vietnam, Malaysia, Java, and Bali. In Borobudur, the ninth-century bas-reliefs depict early records of people giving body massages, which is part of traditional medical practice. The importance of traditional herbs and medicines is also beginning to achieve wider global recognition. It is, therefore, not surprising that Anita Roddick, founder of Body Shop, spent many years in Indonesia researching its indigenous health and beauty products (Beers 2001, 39).

For the Western sojourner in the region, the biggest challenge has been fighting malaria. This was a plague that killed possibly hundreds of thousands of Western sojourners over the centuries. It culminated in the eighteenth century in epidemic proportions in Dutch Batavia, when the city became known as the white man's graveyard (Savage 1984). The Dutch had built Batavia modeled on their own cities with many canals, but these became breeding grounds for mosquitoes. But the knowledge that the mosquito was the cause of malaria was only discovered in the late nineteenth century, although already in Singapore in the mid-nineteenth century there was speculation of a correlation between mosquitoes and malaria (Savage 1984). Once such a correlation was ascertained, the science of public health in the region changed rapidly, as major attempts were made to eliminate stagnant water areas, followed later by spreading insecticides on water surfaces to kill the mosquito larvae.

The local adaption and response to malaria had historically been to live upland in the region above 1,000 meters, and deep in the jungles where the mosquito is less of a menace. In prehistoric times, oceanic migrations across to the Pacific Islands and to Madagascar could also have been to find places where malaria was not a problem. In addition, Southeast Asians also use herbal medicine against malaria. Over 40 species from 32 plant families are used by hill-tribal groups (Akha, Karen, Lisu, Hmong, and Lahu) to treat malaria (Anderson 1993, 134). Just as Western medicines use the bitter drug "quinine" to treat malaria, tribal groups in Southeast Asia also use bitter tasting medicines that suggests that the presence of alkaloids may have therapeutic value against malaria (Anderson 1993, 134).

LESSONS FOR ENVIRONMENTAL EDUCATION: UNDERSTANDING INDIGENOUS RELATIONSHIPS TO NATURE

In the developing global dialogue on environment, ecology, and sustainability, there is growing new respect for the worldviews, folk science, and sustainable living mechanisms of indigenous communities in Southeast Asia and elsewhere in the world. The indigenous knowledge, wisdom, and worldviews are resonating in the deep ecology movements (Devall and Sessions 1985) and in the resurrection of indigenous religions such as Neopaganism as a New Age religion today (Davy 2007; Pike 2004; Shnirelman 2002), and also in the development of environmental ethics (Peterson 2001; Leopold 1977; Callicott 1994). Ironically, the rise of Neopaganists is precisely due to their reverence for Nature as contrasted to Christianity that is seen as a religion that exploits, marginalizes, and denigrates Nature (Davy 2007 and Shnirelman 2002).

The relevance of indigenous knowledge and folk science in reframing the ecological and sustainable living discourses of environmental education lies in the fact that, contrary to Arnold Toynbee's (1976, 14) depiction of modern man's appraisals of the biosphere from the "outside" as both "a spectator and a censor," indigenous communities of Southeast Asia viewed nature and the biosphere very much as participants, insiders, and integral components of a spiritual community.[3]

The *relational* view of Nature is common to Asian religions (Buddhism, Taoism) according to Anna Peterson (2001). By this she means that such religions are embedded in the concept of the "relational self" defined by natural and social relationships, which is a radical alternative to the *individualism* in Western culture and practices (Peterson 2001, 93–94). This relational worldview has made indigenous communities base their interactions with Nature and the environment on concepts of "equity," "respect," "reciprocity," and "reverence" that shape their consumption, commodification, and utilization of the environment.[4]

David Orr writes:

> Ecological education is not just about biology, it is equally about the deeper causes of biotic impoverishment, which have to do in one way or another with political behavior, institutions, and philosophies. Conservation biology is a dialogue between science and political action. (Orr 2004, 72)

The quote is a reminder that environmental education must be embedded in a constant dialogue between the humanities and science and that there is no place for the "culture wars" in such an education. Indeed good environmental science is a prerequisite for reliable social science and humanities discourses and societal applications. And good social science provides an understanding especially in the Anthropocene Age for understanding the processes of environmental change and landscape transformations. Indeed historians of the region have been slow in factoring environmental hazards and climate change impacts in their analysis of the region's expanding and collapsing empires, cities, and civilizations. Victor Lieberman's (2003) recent history of mainland Southeast Asia, entitled *Strange Parallels; Southeast Asia in Global Context, c. 800–1830,* is probably the first comprehensive history of the region to factor in climate change consequences in explaining the rise and fall of civilizations in the region and beyond. Specifically, he draws attention to the El Niño factor in explaining agricultural growth between c. 1000 and 1300 in Burma, Cambodia, and southern China, and droughts and desiccation from c. 1400 to the sixteenth centuries that explains the exhaustion of agriculture and food security. In his view he notes, "both Pagan and Angkor succumbed to a combination of ecological strains in the core, climate deterioration, maritime shifts, and Mongol-assisted Tai incursions" (Lieberman 2003, 242).

In a world with a wide diversity of environments, ecosystems, and natural resources, we must accept that the 196 countries, or "imagined communities" as Benedict Anderson (1991) puts it, do not have monopoly of the wisdom for sustainable living. There are the other 8,600 indigenous communities in the world that also have claim to this wisdom, and we should not allow the tyranny of the state to dictate how communities should coexist and live sustainably with their ecosystems. While imperial rulers conquered territories to control peoples, this obsession with territory and territoriality is certainly the wrong lessons for sustainable living. We need to move the ecosystem perspective into the center of political and economic dialogues and environmental educational policies for sustainable living.

The important lessons for environmental education lie in finding new perspectives and opening up multicultural dialogues to common-pool resources, ecological principles, a relational approach to nature, and a quality of living that is less materialistic, less consumer driven, and less hinged on modernization outcomes. The ecological perspective is pertinent because it provides the ideological undergirding of environmental education that Manuel Castells (2004, 170) views as the "theory" of environmentalism, Andrew Dobson (2001, 1–2) dubs as "ecologism," and Thomas Berry (1999, 84) refers to as "functional cosmology." These perspectives and worldviews are very much reflected in the cognition of indigenous communities, but still remain the terra incognitae of the power elites in the global community.

END NOTES

1. However, from the fifteenth century onward these changes took place increasingly between literate and illiterate peoples who Eric Wolf (1982) refers to as "people without history." Although Wolf (1982, 385) does not address the contributions of "primitives" to the fund of Western knowledge, he argues that the global expansion of the Western world constituted "*their* history as well."

2. Central to this academic discourse and political action are finding cultural and societal adaptive prescriptions and mitigation efforts. While global leaders and nation-state political strategists are engaged in an intellectual enterprise of territorial and marine conquest, economic dominance, cultural imperialism, and military competition, they need to turn their attention to the conquest of their mental realms. The competition and exploits are not in the biosphere per se but in the mental-sphere or what Pierre Teilhard de Chardin (1964, 137) calls "noosphere," the "thinking envelope of the Earth." Or reflected in another way, the understanding of our world is a product of *cosmogenesis*, as a world coming into being through transformations of "lesser to a great order of complexity and from a lesser to great consciousness" (Berry 1999, 26). Put in explorers' parlance, our leaders, politicians, bureaucrats, and entrepreneurs need to explore and conquer their own mental terra incognitae and the noosphere so that they are better informed about climate change, ecological perspectives, cosmogenesis, and human-nature relations.[3] The spiritualization of Nature provides indigenous communities a sense of security, comfort, and oneness with their environment. The most apt way of seeing this is using the definition of religion as "a cognitive and normative structure that makes it possible for man to feel 'at home' in the universe," given by Peter Berger et al. (1973, 79). Hence, unlike modern individuals and communities that have severed their magical, mystical, and religious links with Nature and hence suffer "*a deepening condition of homelessness*" (Berger et al. 1973, 82), the indigenous communities of the region despite their lack of a sedentary habitat never feel a sense of homelessness because they are at home

within their communities, their existential ecosystems, and in the cosmos.[4] This integration of Nature and self is one of the reasons why communities in Southeast Asia could not develop a more scientific enquiry into specific elements of Nature. In short, they could not separate the elements of Nature from self, spirits, ancestors, and environment. Amongst the region's indigenous groups, the relational process is continually resurrected in six symbolic themes: the journey (migration; *Merantau*), the cosmic center (the navel or *pusat*); tripartite cosmos (upper, middle, and lower worlds), the Great Mother (Mother Earth and Father Sky), the Cosmic Tree (or Tree of Life), and Death-Rebirth transformations. In Thailand, for example, Buddhism, Brahmanism, and animism often overlap and reinforce each other in a state of peaceful coexistence. Generally, these relationships are expressed in ceremonies and festivals that give the Thai villager a sense of "psychological securities" (Klausner 1993, 39).

BIBLIOGRAPHY

Anderson, Benedict. 1991. *Imagined Communities*. London: Verso.

Anderson, Edward. 1993. *Plants and People of the Golden Triangle: Ethnobotany of the Hill Tribes of Northern Thailand*. Chiang Mai: Silkworm Books.

Beers, Susan-Jane. 2001. *Jamu: The Ancient Indonesian Art of Herbal Healing*. Hong Kong: Periplus.

Berger, Peter, Brigitte Berger, and Hansfried Kellner. 1973. *The Homeless Mind: Modernization and Consciousness*. New York: Random House.

Berry, Thomas. 1999. *The Great Work: Our Way into the Future*. New York: Bell Tower.

Blaut, J. M. 1993. *The Colonizer's Model of the World: Geographical Diffusionism and Eurocentric History*. New York and London: Guilford Press.

Callicott, Baird J. 1994. *Earth's Insights: A Muilticultural Survey of Ecological Ethics from the Mediterranean Basin to the Australian Outback*. Berkeley and Los Angeles: University of California Press.

Castells, Manuel. 2004. *The Power of Identity*. Oxford: Blackwell Publishing.

Chanda, Nayan. 2007. *Bound Together: How Traders, Preachers, Adventurers, and Warriors Shaped Globalization*. New Haven, CT, and London: Yale University Press.

Conklin, H. C. 1957. "Hanunoo Agriculture." *F.A.O. Forestry Development Paper* 2. Rome: FAO.

Cosgrove, Denis. 2008. *Geography & Vision: Seeing, Imagining and Representing the World*. London: I. B. Tauris.

Dallmayr, Fred. 2002. *Dialogue among Civilizations: Some Exemplary Voices*. New York: Palgrave Macmillan.

Davy, Barbara Jane. 2007. *Introduction to Pagan Studies*. Lanham, MD: AltaMira Press.

Devall, Bill, and George Sessions. 1985. *Deep Ecology*. Layton: Peregrine Smith Books.

Dobson, Andrew. 2001. *Green Political Thought*. London and New York: Routledge.

Dove, Michael. 1990. "Introduction: Traditional Culture and Development in Contemporary Indonesia." In *The Real and Imagined Role of Culture in Development*, edited by M. R. Dove, pp. 1–37. Honolulu: University of Hawaii Press.

Driver, Felix. 2001. *Geography Militant: Cultures of Exploration and Empire.* Oxford: Blackwell Publishers.

Fahn, James David. 2003, *A Land on Fire.* Bangkok: Silkworm Books.

Forsyth, Tim, and Andrew Walker. 2008. *Forest Guardians and Forest Destroyers.* Chiang Mai: Silkworm Books.

Frank, Andre Gunder. 1998. *ReOrient: Global Economy in the Asian Age.* Berkeley: University of California Press.

Garnier, de-Lagrée Francis. 1885. *Voyages D'exploration en Indo-Chine Effectue Par Une Commission Francaise.* Paris: Librarie Hachette et Cie.

Geertz, Clifford. 1980. *Negara: The Theatre State in Nineteenth-Century Bali.* Princeton, NJ: Princeton University Press.

———. 1971. *Agricultural Involution: The Processes of Ecological Change in Indonesia.* Berkeley: University of California Press.

Gelber, Harry G. 2007. *The Dragon and the Foreign Devils: China and the World, 1100 BC to the Present.* London: Bloomsbury.

Groeneveldt, W. P. 1876. "The Expedition of the Mongols against Java in 1923, A.D." *The China Review* 4, no. 4: 216–254.

Hardin, Garrett. 1980. *Promethean Ethics: Living with Death, Competition and Triage.* Seattle: University of Washington Press.

———. 1968. "The Tragedy of the Commons." *Science* 162: 1243–1248.

Headley, John M. 2008. *The Europeanization of the World: On the Origins of Human Rights and Democracy.* Princeton, NJ, and Oxford: Princeton University Press.

Hobson, John M. 2004. *The Eastern Origins of Western Civilisation.* New York: Cambridge University Press.

Howell, Signe. 1989. *Society and Cosmos: Chewong of Peninsula Malaysia.* Chicago: University of Chicago Press.

Klausner, William J. 1993. *Reflections on Thai Culture.* Bangkok: The Siam Society.

Lan, Tuyet Chu. 2002. "An Introduction to the History of Traditional Medicine and Pharmaceutics in Vietnam." In *Historical Perspectives in East Asian Science, Technology and Medicine,* edited by Alan, K. L. Chan, Gregory K. Clancy, and Loy Hui-Chieh, pp. 264–284. Singapore: Singapore University Press and World Scientific Publishing Co.

Lee, Sung Kyu. 2002. "Traditional East Asian Views of Nature Revisited." In *Historical Perspectives in East Asian Science, Technology and Medicine,* edited by Alan K. L. Chan, Gregory K. Clancy, and Loy Hui-Chieh, pp. 60–65. Singapore: Singapore University Press and World Scientific Publishing Co.

Leopold, Aldo. 1977. *A Sand County Almanac: With Essays on Conservation from Round River.* New York: Ballantine Books.

Li, Hui-Lin. 1979. *Nan-fang Ts'ao-mu Chuang: A Fourth Century Flora of Southeast Asia.* Hong Kong: The Chinese University Press.

Lieberman, Victor. 2003. *Strange Parallels Southeast Asia in Global Context, c. 800–1830, Volume 1 Integration on the Mainland.* Cambridge: Cambridge University Press.

Loubere, Simon de la. 1693. *A New Historical Relation of the Kingdom of Siam,* vols. 1 and 2. Translated by A. P. Gen. R. S. S. London: F. L. for Tho. Horne at the Royal Exchange, Francis Saunders at the New Exchange, and Tho. Bennet at the Half-Moon in St. Pauls Church-yard.

Menzies, Gavin. 2003. *1421: The Year China Discovered the World.* London: Bantam Books.

Nguyen, Dien Xuan. 2002. "Ancient Vietnamese Manuscripts and Printed Books Related to Science, Medicine and Technology (Inventory, Classification and Preliminary Assessment)." In *Historical Perspectives in East Asian Science, Technology and Medicine,* edited by Alan K. L. Chan, Gregory K. Clancy, and Loy Hui-Chieh, pp. 547–554. Singapore: Singapore University Press and World Scientific Publishing Co.

Oppenheimer, Stephen. 1999. *Eden in the East: The Drowned Continent of Southeast Asia.* London: Phoenix.

Orr, David W. 2004. *Earth in Mind: On Education, Environment, and the Human Prospect.* Washington, DC: Island Press.

Osborne, Roger. 2007. *Civilization: A New History of the Western World.* London: Pimlico.

Peterson, Anna L. 2001. *Being Human: Ethics, Environment, and Our Place in the World.* Berkeley: University of California Press.

Pike, Sarah M. 2004. *New Age and Neopagan Religions in America.* New York: Columbia University Press.

Pollock, Sheldon. 2006. *The Language of the Gods in the World of Men: Sanskrit, Culture, and Power in Premodern India.* Berkeley: University of California Press.

Reynolds, C. J. 1995. "A New Look at Old Southeast Asia." *The Journal of Asian Studies* 54, no. 2: 419–446.

Rosaldo, Renato. 1994. *Ilongot Headhunting 1883–1974: A Study in Society and History.* Stanford, CA: Stanford University Press.

Rumpf, Georg Eberhard. 1705. *D'Amboinsche rariteitkamer: behelzende eene beschryvinge van allerhande zoo weeke als harde schaalvisschen, te weete raare krabben, kreeften,en diergelykezeedieren, als mede allerhande hoorntjes en schulpen, die men in d'Amboinsche Zee vindt: daar benevens zommige mineraalen, gesteenten, en soorten van aarde, die in d'Amboinsche, en zommige omleggende eilanden gevonden worden.* Amsterdam: Gedrukt by F. Halma.

Sachs, Jeffrey. 2005. *The End of Poverty: How We Can Make It Happen in Our Lifetime.* London: Penguin Books.

Said, Edward W. 1979. *Orientalism.* New York: Vintage Books.

Sauer, Carl O. 1969. *Agricultural Origins and Dispersals.* Cambridge: The MIT Press.

Savage, Victor R. 2009. "A Question of Space: From Aterritorial Communities to Colonized States in Southeast Asia." Unpublished Conference Paper.

———. 1984. *Western Impressions of Nature and Landscape in Southeast Asia.* Singapore: University of Singapore Press.

Shnirelman, V. A. 2002. "'Christians! Go Home': A Revival of Neo-paganism between the Baltic Sea and Transcaucasia (An Overview)." *Journal of Contemporary Religion* 17, no. 2: 197–211.

Solheim, W. G. II. 1970. "Northern Thailand, Southeast Asia and World Prehistory." *Asia Perspectives* 13: 145–162.

Sonnerat, Pierre. 1776. *Voyage a la Nouvelle Guinée, dans lequel on trouvé; a description des lieux. Des Pnservations physiques & morales, & des details relatives a l'Histoire Naturelle dans le Regne Animal & le Rene vegetal.* Paris: Chez Ruault, Libraire.

Spyer, Patricia. 2000. *The Memory of Trade: Modernity's Entanglements on an Eastern Indonesian Island.* Durham, NC, and London: Duke University Press.

Stark, Rodney. 2005. *The Victory of Reason: How Christianity Led to Freedom, Capitalism, and Western Success.* New York: Random House.

Steadman, John M. 1969. *The Myth of Asia.* New York: Simon & Schuster.

Teilhard de Chardin, Pierre. 1964. *The Future of Man.* Translated by Norman Denny. New York and Evanston: Harper Torchbooks.

Toynbee, Arnold. 1976. *Mankind and Mother Earth.* New York: Oxford University Press.

Tsing, Anna L. 1993. *In the Realm of the Diamond Queen: Marginality in an Out-of-the Way Place.* Princeton, NJ: Princeton University Press.

Valentijn, François. 1724–1726. *Oud En Nieuw Oost-Indiën.* Dordrecht: J. van Braam and Amsterdam: G. O. de Linden.

van Leur, J. C. 1955. *Indonesian Trade and Society: Essays in Asian Social and Economic History.* The Hague and Bandung: W. van Hoeve.

Wallerstein, Immanuel. 1989. *The Modern World-System, Vol. 3, The Second Era of Great Expansion of the Capitalist World-Economy 1730–1840s.* New York: Academic Press.

———. 1980. *The Modern World-System, Vol. 2, Mercantilism and the Consolidation of the European World-Economy 1600–1750.* New York: Academic Press.

———. 1974. *The Modern World-System, Vol. 1, Capitalist Agriculture and the Origins of the European World-Economy in the Sixteenth Century.* New York: Academic Books.

Wheatley, Paul. 1961. *The Golden Khersonese.* Kuala Lumpur: University of Malaya Press.

Wolf, Eric R. 1982. *Europe and the People without History.* Berkeley: University of California Press.

Wolters, O. W. 1999. *History, Culture, and Region in Southeast Asian Perspectives.* Ithaca, NY, and Singapore: Southeast Asia Program Publications and Institute of Southeast Asian Studies.

Wong, David S. Y. 1977. *Tenure and Land Dealings in the Malay States.* Singapore: Singapore University Press.

Wyatt, David K. 2002. *Siam in Mind.* Chiang Mai: Silkworm Books.

Yoon, Hong-key. 2006. *The Culture of Fengshui in Korea: An Exploration of East Asian Geomancy.* Lanham: Lexington Books.

Yukio, Ōhashi. 2002. "Originality and Dependence of Traditional Astronomies in the East." In *Historical Perspectives in East Asian Science, Technology and Medicine,* edited by Alan K. L. Chan, Gregory K. Clancy, and Loy Hui-Chieh, pp. 394–405. Singapore: Singapore University Press and World Scientific Publishing Co.

AUTHOR INDEX

SUBJECT INDEX